We Are Stardust

Robert Fleck

We Are Stardust

Stellar Evolution and Our Cosmic Connection

 Springer

Robert Fleck
Emeritus Professor of Physics and Astronomy
Department of Physical Sciences
Embry-Riddle Aeronautical University
Daytona Beach, FL, USA

ISBN 978-3-031-67274-3   ISBN 978-3-031-67275-0   (eBook)
https://doi.org/10.1007/978-3-031-67275-0

This Springer imprint is published by the registered company Springer Nature Switzerland AG
The registered company address is: Gewerbestrasse 11, 6330 Cham, Switzerland

If disposing of this product, please recycle the paper.

We are stardust
Billion-year-old carbon…

      —Singer-songwriter Joni Mitchell, "Woodstock" (1969)

Stars have a life cycle much like animals. They get born, they grow, they go through a definite internal development, and finally they die, to give back the material of which they are made, so that new stars may live.

   —Hans Bethe concluding his 1967 physics Nobel acceptance speech

… we have found it possible to explain, in a general way, the abundances of practically all the isotopes of the elements from hydrogen through uranium by synthesis in stars and supernovae.

   —F. Hoyle, William A. Fowler, G. R. Burbidge, and E. M. Burbidge,   "Origin of the Elements in Stars," *Science* **124** (1956, p. 611)

We are made of stardust. Every atom of every element in your body except for hydrogen [a product of the Big Bang 13.8 billion years ago] *has been manufactured inside stars, scattered across the Universe in great stellar explosions, and recycled to become part of you.*

       —John Gribbon, *Stardust* (2000, p. 1)

All of the rocky and metallic material we stand on, the iron in our blood, the calcium in our teeth, the carbon in our genes were produced billions of years ago in the interior of a red-giant star. We are made of star-stuff.

—Carl Sagan, *The Cosmic Connection* (1973, pp. 189–190)

… we are the local embodiment of a Cosmos grown to self-awareness. We have begun to contemplate our origins: starstuff pondering the stars, organized assemblages of ten billion billion billion atoms; tracing the long journey by which, here at least, consciousness arose… Our obligation to survive is owed not just to ourselves but also to that Cosmos, ancient and vast, from which we sprang.

—Carl Sagan, *Cosmos* (1980, p. 345)

You are a child of the Universe…

—Max Ehrmann, *Desiderata* (1927)

For you are dust and to dust you shall return.

—Genesis 3:19

Acknowledgements

It is a pleasure to acknowledge, once again, the encouragement and support of Angela Lahee, physics executive editor for Springer Publishing. And once again, I thank a lifetime of students for working with me to understand the life cycle of stars, the very essence of our cosmic connection. It has been a pleasure and a privilege to have spent a lifetime in the classroom with all of them. I thank Joni Mitchell for long ago opening my eyes—and ears—to our *real* cosmic connection through her beautiful music, true "music of the spheres." There will never be another like her.

I dedicate this book to all the children of the stars—all of us star-stuff star folk—riding together on this rock we call Earth around the Sun, our star in the Cosmos, with the hope that, appreciating our common cosmic connection, we will all make the ride safer and better for everyone, for the rock itself and all the life it supports, and for all who will follow on this cosmic journey.

... still on the Beach in Daytona   Robert Fleck
USA
June 2024

Contents

About the Author

Robert Fleck is Emeritus Professor of Physics and Astronomy in the Department of Physical Sciences at Embry-Riddle Aeronautical University in Daytona Beach, Florida, where for four decades he developed and taught a large number and a wide variety of undergraduate and graduate courses in physics, astronomy, general science, and history of science. For inspiring his students with his passion and enthusiasm for teaching and lifelong learning, he received the University Outstanding Teaching Award in 2000 and 2015, as well as over a dozen faculty appreciation awards from graduating senior classes. Professor Fleck is a NASA and National Science Foundation supported star and planet formation theorist; he has published in a wide variety of disciplines, including physics and astronomy and the history of science, and he has been a Visiting Scientist at the National Radio Astronomy Observatory and a Perren Visiting Fellow at the University of London. He also pioneered Embry-Riddle's study abroad program, teaching classes in England, France, Italy, and Greece, and he has completed two book-length manuscripts titled *The Evolution of Scientific Thought: A Cultural History of Western Science from the Paleolithic to the Present* and *Art History as Science History: Picturing the History of Science from the Paleolithic to the Present*. His recently published book is *Entropy and the Second Law of Thermodynamics …or Why Things Tend to Go Wrong and Seem to Get Worse*. When not reading or writing, he enjoys swimming, surfing, cycling, and traveling.

1

Introduction

Summary In the days before books and television, before computers and cell phones, before Facebook and Instagram, our ancient ancestors watched and wondered about their place in the Universe and composed stories—myths—to organize their experiences, many of which expound some aspect of cosmic order. Modern science has shown that, in a very real and profound sense, we are intimately connected to the Cosmos: we—and all we see around us—are a natural product of the workings and wonders of the Universe. This introductory chapter summarizes this book's purpose to help the reader understand and appreciate this cosmic connection, that we are, in fact, tied directly to distant events spread across the Universe in space and time reaching back to the beginning, back to the Big Bang, and continuing through the birth and death of successive generations of stars. We are stardust—in a very real sense, children of the stars—star folk made from chemical elements ("star stuff") created by nuclear reactions in stellar furnaces and distributed throughout the Galaxy during the various stages of stellar evolution. Whether or not the stars above us govern our conditions (through astrological influences, as suggested, for example, in William Shakespeare's *King Lear*), they have certainly helped *create* our conditions—and, significantly, since these processes occur throughout the Galaxy, the conditions for life throughout the Universe. It is hoped that the reader will leave, awed by the power and beauty of this cosmic perspective, with a better understanding and appreciation of our true cosmic connection.

© The Author(s), under exclusive license to Springer Nature
Switzerland AG 2024
R. Fleck, *We Are Stardust*, https://doi.org/10.1007/978-3-031-67275-0_1

1

We are stardust
Billion-year-old carbon . . . .
                                —Singer-songwriter Joni Mitchell, "Woodstock" (1969)

... if you ask about the origin of the [chemical] *elements ... the answer is that*
hydrogen and helium are left from an early, hot, dense phase in the life of the
Universe (called the Big Bang); all the rest were made by the stars.
                                —astronomer Virginia Trimble, "The Origin and Evolution
                                                of the Chemical Elements" [1, p. 63]

In *Cosmos*, a hugely popular story of the Universe and our place in it,[1] the late astronomer Carl Sagan finds "something curious about the national flags of the planet Earth": almost half display astronomical symbols such as stars (one for many countries including Russia and Israel, five for China, fifty for the United States with stars featured on nearly half the state flags as well), the Sun (e.g., Japan and Argentina), the crescent Moon (Turkey and many of the Islamic states), the celestial sphere (Brazil), various constellations (e.g., Australia and New Zealand), as well as a variety of cosmological symbols (e.g., India and South Korea). "The phenomenon," Sagan points out [2, p. 50],

is transcultural, nonsectarian, worldwide. It is also not restricted to our time: Sumerian cylinder seals from the third millennium BC and Taoist flags in prerevolutionary China displayed constellations. Nations, I do not doubt, wish to embrace something of the power and credibility of the heavens. *We seek a connection with the Cosmos* [emphasis added]. We want to count in the grand scale of things. And it turns out we *are* connected—not in the personal, small-scale unimaginative fashion that the astrologers pretend, but in the deepest ways, involving the origin of matter....

Indeed, we have always looked to the heavens for inspiration and answers; there is something uplifting and comforting about looking up to the sky at

---

[1] Although some parts are, not surprisingly, a bit outdated (thus does science advance), Carl Sagan's *Cosmos*, the book that accompanied the 1980 PBS documentary series of the same title, is still well worth a read. I remember reading it when it first appeared in the early years of my career shortly after leaving university with a Ph.D. in astronomy. Of course, I found the science fascinating and, Ph.D. notwithstanding, I learned a lot of astronomy. But even more than that, what really impressed me was the breadth of Sagan's knowledge reaching far outside the confines of astronomy, and his almost superhuman ability to share that knowledge and enthusiasm. His book opened my eyes to the world outside astronomy, inspiring me to learn all that I can about all that I can. When I think of genius, Sagan still sits atop all others in ability to communicate science effectively and enthusiastically, a rare and much needed talent especially today (see, e.g., Brandon Brown's *Sharing Our Science: How to Write and Speak STEM*, MIT Press, 2023; others I place in the genius category include The Beatles for music, Robin Williams for humor, and the 1972 Miami Dolphins for football). Give *Cosmos* a read. You won't be disappointed.

night [3–5].[2] After providing for the food and shelter necessary for survival—the essentials biologists call the "four Fs" of life: feeding, fighting, fleeing, and mating—our ancient ancestors, in the days before books and television, before computers and cell phones, before Facebook and Instagram, watched and wondered about their place in the Universe. Despite the proliferation of modern distractions, we still wonder, and we still ask ourselves fundamental questions regarding our origin, purpose, and destiny (Fig. 1.1).[3] Different cultures ask different questions seeking different kinds of knowledge and often arrive at answers, different kinds of "truths." As we became better at figuring things out, our answers to these questions evolved through increasing stages of sophistication, progressing from primitive mythic narrative to recognizably mature scientific responses.[4]

As Sagan reminded us [2, pp. xii, 318],

... science has found not only that the universe has a reeling and ecstatic grandeur, not only that it is accessible to human understanding, but also that we are, in a very real and profound sense, a part of the Cosmos, born from it, our fate deeply connected with it....

Something in us recognizes the Cosmos as home. We are made of stellar ash. Our origin and evolution have been tied to distant events....

As the ancient mythmakers knew, we are the children equally of the sky and the Earth....

---

[2] Uplifting and more. "It is one of the ironies of history," historian of science John North concludes in his historical survey of astronomy and cosmology [6, p. 623], "that the study of such a vast and impersonal subject matter [i.e., astronomy] should from beginning to end have been so intimately bound up with principles of human nature." Most early societies were organized around the sky, often associating their gods with celestial bodies, and most of us today continue to be fascinated with the study of astronomy.

[3] Consider the "rationale" for a recently advertised symposium sponsored by the International Astronomical Union titled "(Toward) Discovery of Life Beyond Earth and its Impact":

From our origins, humans have been inspired by pinpoints of light in the night sky. They cause us to wonder about our existence. Who are we? What are we doing here? Where did we come from? And, where are we going?. . . Despite impressive investment and activity in space exploration over the years, the question remains unanswered.

Gauguin's questions (see Fig. 1.1) still strike at the heart of modern science.

[4] See, for example, J. Norman Lockyer's 1894 *The Dawn of Astronomy* [7], a pioneering study of the interplay of archaeology and astronomy in the ancient world. Lockyer identified three advancing stages in the historical development of astronomy: the mythical, the practical, and the intellectual. Although Lockyer confined his attention to astronomy in particular, one could easily assign these developmental stages to science in general. The developmental stages in our understanding of our connection to the Cosmos are examined in subsequent chapters.

Fig. 1.1 French post-Impressionist artist Paul Gauguin's monumental (measuring nearly five feet by just over 12 feet in size) 1897 *D'où Venons Nous/Que Sommes Nous/Où Allons Nous* (*Where do we come from? What are we? Where are we going?*), now in the Boston Museum of Fine Arts. The title in French verse appears in the upper-left-hand corner of the painting which was made during the artist's Polynesian period. Ever since we became conscious of ourselves and the world around us, we have asked "ultimate" questions such as these regarding our origin, purpose, and destiny. We know from a surviving manuscript in which he extolled his philosophical musings about human destiny, that Gauguin considered the role that new scientific knowledge might play in resolving such questions. In any case, it is significant that when Gauguin posed them, science was not in a position to provide answers; one would have had to consult philosophers or theologians. Today, however, more than a century later, science has come a long way in answering these most profound of questions that still lie at the heart of philosophy and modern science—even if, to the dismay of author John Gribbin, "hardly anybody outside a small circle of scientific specialists seems to have noticed" that "[w]e have answered the biggest question of them all—where do we come from?" [8, p. 177], Gauguin's leading question here. Even more profoundly, the late English theoretical physicist and cosmologist Stephen Hawking wanted to know *Why* are we? (*Wikimedia Commons, public domain*)

As a star and planet formation theorist, I have for a long time wanted to write this book about our *real* cosmic connection, our connection with the stars, not the presumed numinous connection suggested by the ancient pseudoscience of astrology, but rather a most intimate connection now recognized by science: the fact that, as Joni Mitchell sings in her song "Woodstock," "we are stardust," all of us—bones, blood, and brains—children of the stars, star folk all made from the same stuff—star stuff—all brought into existence with the birth of the Universe and the evolution—the change in the structure and composition from birth to death—of stars. The oxygen in the air we breathe and in the water that makes up most of our body mass, the carbon so essential to all organic material, the nitrogen in our amino and nucleic acids that are the building blocks of our proteins and genetic material, the calcium in our bones and teeth, the iron in our blood—*all of us—and all that we see around us are a natural product of the workings and wonders of the Universe* (Fig. 1.2).

Fig. 1.2 *Arizona*, painted in 1979 by the American artist Ray Swanson (1937–2004). While it may not have been the primary intent of the artist, Swanson's painting (and several others such as *Monument Valley Lady* and *Eskimo Lady and Her Land*) are vivid illustrations of our cosmic connection, that we and everything around us share a common origin deep in the history of the Universe, a view that resonates with that of Indigenous Peoples, who believe that *all* things are connected, that we are people equally of Earth and Sky. Note the remarkable similarities in color and form of the age-worn Navaho woman and the eroded ancient rock formations of Arizona's Grand Canyon. (*Courtesy of the Ray Swanson family estate. Used with permission*)

Astrophysicist John Gribbin makes it very clear in the final sentence of his book: "We are made of stardust because we are a natural consequence of the existence of stars," adding, as a result, addressing what is perhaps the only other ultimate cosmic question—the question of life elsewhere in the Universe—that "from this perspective it is impossible to believe that we are alone in the Universe" [8, p. 187; recall Note 3].

Our story begins where the Universe itself began: with the Big Bang some 13.8 billion years ago when, during the first three minutes in the history of the Universe (to borrow from the title of Physics Nobel laureate Steven Weinberg's account of the origin of the Universe [9]), all of the hydrogen and most of the helium—by far the most abundant elements in the Universe which is roughly three-quarters hydrogen and one-quarter helium by mass—formed from a cooling plasma of protons, neutrons, and electrons.[5] We then

---

[5] All the stuff ("stof" is Danish for "matter") that we see all around us (including us) all the way out to the farthest reaches of the Universe—what we call *baryonic matter* because it consists primarily of the two most common baryons, positively charged *protons* and electrically neutral *neutrons*, the basic building blocks of ordinary matter—amounts to just under 5% of the total mass-energy of the Universe. (Recall that in his theory of relativity, Albert Einstein demonstrated the equivalence of mass (m) and energy (E) which are related by arguably the most famous equation in science, $E = mc^2$, c denoting the speed of light—at 186,000 miles per second a *very* big number—nearly as fast as the speed of *life*!—so even a little mass contains a lot of energy.) The word "baryon" comes

trace the life cycle of the stars from birth to death—*stellar evolution*—highlighting the synthesis in self-gravitating nuclear fusion reactors called stars, of the heavier chemical elements so essential to life, along the way touching on many of the hot topics in astrophysics today including exoplanets, supernovae, pulsars, black holes, white dwarfs, and life in the Universe, all with a minimum of mathematics (developed mostly in Notes that, in any case, can be skipped over). "You are a child of the universe" the American author Max Ehrmann reminds readers of his prose poem, *Desiderata*, written a century ago, "stardust, billion-year-old carbon," Joni Mitchell sings. Humans—and all life on Earth sharing a common carbon chemistry—are, as the film *Star Trek: The Motion Picture* reminded us, "carbon-based units."[6] Indeed, *life as we know it is an inevitable consequence of the life cycle of the stars.*

---

from the Greek "barus" meaning "heavy": the other component of ordinary matter, the electron, carrying a negative electric charge equal in magnitude to the positive charge of the proton, is some two-thousand times lighter than the proton and neutron which have nearly the same mass—nearly the same, but, as we'll see, importantly not exactly the same: the neutron is slightly more massive than the proton. (The *electron* is the most common type of *lepton* [Greek, "small"]; the nearly massive *neutrino* is also a lepton and is the most abundant elementary particle with mass in the Universe, perhaps more abundant than the *photon*, the particle of light—even if it has no electric charge and so little mass that it is so difficult to detect.) And so, for us, there are only three primary constituents of ordinary matter—one less than the four of ancient Greece (earth, water, air, and fire).

Amazingly, most of the stuff in the Universe is dark (invisible to the eye), and although we have no idea what it is (if you, the reader, have any ideas, please share them with me: I'll win the Nobel Prize and I'll buy you a beer), we do know that it's about three quarters *dark energy*, mysterious stuff that, among other things, drives the expansion of the Universe, and about one quarter *dark matter*, stuff we know is out there because of the way it interacts gravitationally with the ordinary matter we can see, each one merely a label for our ignorance. Astronomer John Barrow refers to this as "the final Copernican twist to our status in the material universe. Not only are we not at the center of the universe; we are not even made out of the predominant form of matter in the universe" [10, p. 74]. (Interestingly, but nonetheless coincidentally, the proportion of dark energy to dark matter mirrors that of hydrogen to helium, the bulk of ordinary matter in the Universe.) Dark matter was once called "missing mass" because, although it reveals itself dynamically, we can't see it so it is, in that sense, missing. Rather like a nighttime map of Earth from space doesn't faithfully represent population density, mirroring instead the distribution of wealth, a map of the light distribution in the Universe is not a good guide of the distribution of matter. I have to admit that I'm totally in the dark over all of this dark stuff (although I often wonder if dark matter might be lost airline luggage), but a lot of clever people are looking into the matter, so to speak—and actively looking for the stuff. Not an easy task, as you can imagine. Stay tuned.

[6] Carbon is the fourth most abundant element in the Universe after hydrogen, helium, and oxygen. It's the basis of organic chemistry, the chemistry of life. Carbon's ability to form stable bonds with many elements, including itself, allows it to form a large variety of very large and complex molecules required for life processes. All living organisms on Earth contain a total of 550 billion tons of carbon, second only to oxygen, out of about 3.6 trillion tons of biomass [11].

Here is Carl Sagan's succinct one-paragraph account in biblical cadence of the Universe's first 10 billion years [2, pp. 337–338; emphasis added]:

For unknown ages after the explosive outpouring of matter and energy of the Big Bang, the Cosmos was without form. There were no galaxies, no planets, no life. Deep, impenetrable darkness was everywhere, hydrogen atoms in the void. Here and there denser accumulations of gas were imperceptibly growing, globes of matter were condensing—hydrogen raindrops more massive than suns. Within these globes of gas was first kindled the nuclear fire latent in matter. A first generation of stars was born, flooding the Cosmos with light. There were in those times not yet any planets to receive the light, no living creatures to admire the radiance of the heavens. *Deep in the stellar furnaces the alchemy of nuclear fusion created heavy elements, the ashes of hydrogen burning, the atomic building materials of future planets and lifeforms.* Massive stars soon exhausted their stores of nuclear fuel. Rocked by colossal explosions, they returned most of their substance back into the thin gas from which they had once condensed. Here in the dark lush clouds between the stars, new rain-drops made of many elements were forming, later generations of stars being born. Nearby, smaller raindrops grew, bodies far too little to ignite the nuclear fire, droplets in the interstellar mist on their way to form the planets. Among them was a small world of stone and iron, the early Earth.

Eventually, some 13.8 billion years after the Big Bang and 4.5 billion years after Earth formed, the "ash of stellar alchemy" emerged into conscious-ness. Humankind arrived. This is the story developed here, our story, the story of our cosmic connection.[7] It has, as Sagan admits, "the sound of epic myth." But, Sagan continues, "it is simply a description of cosmic evolu-tion as revealed by the science of our time." I hope that you, the reader, will leave, awed by the power and beauty of this cosmic perspective, with a better understanding and appreciation of our true cosmic connection.

To me, as an astrophysicist—as a human being—*our understanding that the stuff we are made of traces its origin to nuclear processes accompanying the Big Bang, and thereafter to billions of years of the birth and death of generation*

---

[7] John Gribbin's *Stardust* has been the only accessible detailed account of our cosmic connection, although, like Sagan's *Cosmos*, it is now somewhat outdated (for example, very recent evidence suggests that many of the heaviest elements, like the silver and gold in your jewelry and the uranium that fuels our nuclear reactors, are produced explosively during neutron star mergers). His differs from mine in that, like Sagan, I have woven historical antecedents and cultural consequences into my story; Gribbin focuses entirely on the science. Surprisingly, despite its profound significance—cosmically and otherwise—this fascinating story of our connection to the stars has largely gone unnoticed: as Gribbin complains [8, p. 177], "hardly anybody outside a small circle of scientific specialists seems to have noticed." Such a beautiful story certainly deserves more attention. *You Are Stardust* by Elin Kelsey (Owlkids Books, Toronto, 2012) shares the same storyline with children, little ones who are naturally curious about everything (would that we all could remain children).

after generation of stars, is one of the most profound and inspiring discoveries ever made, certainly *the* most fundamental and fascinating finding about ourselves and our connection to the Cosmos. This was the most important message I hoped my students would take away from their study of astronomy. I'm not alone in my rhapsodic feelings here: introducing *Carl Sagan's Cosmic Connection: An Extraterrestrial Perspective* with her essay "Carl Sagan: A New Sense of the Sacred," Ann Druyan, Sagan's one-time wife and award-winning producer and director who co-wrote the 1980 PBS documentary series *Cosmos*, admits to a

> soaring spiritual high that is science's overarching revelation—our oneness with the cosmos…. We are starstuff. You, me and everybody. Not the failed clay of a disappointed Creator, but, literally, down to every atom in our bones, the ash of stars [12, p. xxvi].

Like Carl Sagan and many others, I like to think that the world will be a better place if more of us realize this place wasn't put here for us, that we are not the "crown of creation" (to borrow from a Jefferson Airplane 1968 album and song title) but rather part of a larger scheme in an indifferent Cosmos, nothing more—or less—as Sagan reminds us, than "the latest manufactures of the galactic hydrogen industry" [2, pp. 338–339]. Sagan believed that "The deflation of some of our more common conceits [our "naïve self-love"] is one of the practical applications of astronomy" [12, p. xxx]. And I hope that you the reader will, like the famous Nobel Prize-winning physicist Richard Feynman, experience "the pleasure of finding things out" (to borrow from the title of his book [13]; Fig. 1.3), and will appreciate our cosmic connection as an antidote to the feeling of insignificance when contemplating the immensity of the Universe. Besides, as Steven Weinberg confesses in the final sentence of his blow-by-blow account of the Universe's *First Three Minutes* [9, p. 155], "The effort to understand the Universe is one of the very few things that lifts human life a little above the level of farce, and gives it some of the grace of tragedy."

Fig. 1.3 Following Feynman in experiencing "the pleasure of finding things out" in a quest for knowledge—in this case here, wondering about the workings of the Universe and our place in it—a cosmically curious observer peaks through Shakespeare's star-studded "brave o'er hanging firmament" to explore the mysterious Empyrean beyond in this illustration taken from the French astronomer Camille Flammarion's *L'Atmosphère: Météorologie Populaire* (*The Atmosphere: Popular Meteorology*) published in Paris in 1888, a time when we could only imagine our cosmic connection. Today we know we are connected to the Cosmos in the most materially intimate way. (*Wikimedia Commons, public domain*)

References

1 V. Trimble, "The Origin and Evolution of the Chemical Elements," in *Origin and Evolution of the Universe: From Big Bang to Exobiology*, 2nd edn., ed. by M. A. Malkan, B. Zuckerman (World Scientific, Singapore, 2020), pp. 63–97

2 C. Sagan, *Cosmos* (Random House, New York, 1980)

3 E.C. Krupp, *Beyond the Blue Horizon: Myths and Legends of the Sun, Moon, Stars, and Planets* (HarperCollins, New York, 1991)

4 J. Marchant, *The Human Cosmos: Civilization and the Stars* (Dutton, New York, 2020)

5 R. Trotta, *Starborn: How the Stars Made Us (and Who We Would Be Without Them)* (Basic Books, New York, 2023)

6 J. North, *The Norton History of Astronomy and Cosmology* (W. W. Norton and Company, New York, 1994)

7 J. N. Lockyer, *The Dawn of Astronomy.* (Dover Publications, Mineola, NY, 2006; orig. publ. Cassell and Company Ltd., London, Paris & Melbourne, 1894)

8 J. Gribbin, *Stardust* (The Penguin Group, Chatham, UK, 2000)

9 S. Weinberg, *The First Three Minutes: A Modern View of the Origin of the Universe* (New York, Basic Books, 1977; rev. ed. 1993)

10 J. D. Barrow, *The Origin of the Universe* (Basic Books, New York, 1994)

11 M. Y. Marov, "Astronomical and Cosmochemical Aspects of the Life Origin Problem," Astron. Rep. **67**, 764–789 (2023). https://doi.org/10.1134/S1063772923080073

12 C. Sagan, *Carl Sagan's Cosmic Connection: An Extraterrestrial Perspective* (Cambridge University Press, 2000; orig. publ. *The Cosmic Connection*, Doubleday & Company Inc., 1973)

13 R. Feynman, *The Pleasure of Finding Things Out* (Perseus Books, Cambridge, MA, 1999)

2

In the Beginning …

Summary We long to be connected to the Cosmos—witness the pseudo-science of astrology, the belief that celestial events profoundly influence our lives here on Earth, popular even today—and we have always looked to the heavens for inspiration and answers. We watched and wondered about our place in the Universe and composed stories—myths—to organize our experiences, many of which expound some aspect of cosmic order. These stories, attempts to feel more at home in an uncertain and often dangerous Universe, typically involved anthropomorphic sky gods, such as the Sumerian god Anu or the biblical God of Abraham. Native Americans and Indigenous Peoples throughout the world all looked to the sky for answers. We were people equally of Earth and Sky. As we became better at figuring things out, our answers to these questions evolved through increasing stages of sophistication, progressing from primitive mythic narrative through the philosophical speculation of the ancient Greeks to recognizably mature scientific responses. The scientific evidence leading to our current understanding of *the* beginning—of the origin of the Universe itself—is reviewed with a final focus on primordial nucleosynthesis, the synthesis of chemical elements in the early Universe. Within the first few minutes after the Big Bang that brought the Universe into existence 13.8 billion years ago, only hydrogen and helium and a trace of lithium were formed. We have to look, as we will in the next chapter, to the stars to understand the origin of the remaining naturally occurring elements. Therein lies our true cosmic connection.

© The Author(s), under exclusive license to Springer Nature
Switzerland AG 2024
R. Fleck, *We Are Stardust*, https://doi.org/10.1007/978-3-031-67275-0_2

In the beginning the Universe was created. This has made a lot of people very angry and has been widely regarded as a bad move.
> — Douglas Adams, *The Restaurant at the End of the Universe* (1980)[1]

Arthur: *What a Spectacle! Millions and millions of stars. Funny thing, the Universe. Sort of scares the shit out of you, doesn't it?*
Kim: *So where did it all come from, do you think?*
Arthur: *Don't know. Just happened I suppose.*
> — from the 2007 film, *When Did You Last See Your Father?*

I'm astounded by people who want to 'know' the Universe when it's hard enough to find your way around Chinatown.
> — American comedian and actor Woody Allen (b. 1935)

We're good at making light of the most profound of matters. But, all joking aside, since our first thoughts, we've been innately curious about origins—of species (including—indeed, especially—our own), of life, of the world around us, and, reaching out to the grandest of scales, of the Universe itself.[2]

After a brief tour of early cosmologies—attempts to "order" (Greek, *kosmos*; whence, for example, *cos*metics)[3] the world around us—ranging from mythic narratives to the philosophical speculation of the ancient Greeks and the radical revisions of the early modern period, our current understanding of the origin and evolution of the Universe is examined with an eye to the creation of the chemical elements—and hence to our cosmic connection—during the early Universe.

---

[1] Adams had much to say about life, the Universe, and everything from his restaurant at the end of the Universe, one other being that "There is a theory which states that if ever anybody discovers exactly what the Universe is for and why it is here, it will instantly disappear and be replaced by something even more bizarre and inexplicable. There is another theory which states that this has already happened."

[2] Concerning origins, those lucky to have lived through some of the best music ever, may have wondered, along with Barry Mann in 1961,
Who put the bomp in the bomp bah bomp bah bomp?
Who put the ram in the rama lama ding dong?
Who put the bop in the bop shoo bop shoo bop?
Who put the dip in the dip da dip da dip?
Yes, we've always wondered and asked—and sometimes sang—about origins. I'm still wondering where the "bop" came from.

[3] Cosmology addresses the makeup of the Universe, whereas cosmetology is concerned with the universe of makeup.

2.1   Early Cosmologies

From the dawn of history, humankind has invented stories about the origins of our world, and deities that were instrumental in its creation, from the Sumerian god Anu, or Sky Father, to the Greek myths about Gaia being created out of Chaos and the Genesis myths of the Abrahamic religions, which are still believed as literal truths in many societies around the world.

— Jim Al-Khalili, *The World According to Physics* (2022), p. 1

It is the nature of sentient creatures to make sense of the world around them. Their survival depends on it. Of all the life Earth has known, our species, *Homo sapiens* ("wise man"), is the only one to have formulated and recorded its thoughts and impressions—indeed, as far as we know, the only one capable of doing so.

And we didn't waste much time. Archeological evidence suggests that nearly as soon as we arrived, we have wondered about our place in the Universe [1]. At some level, however primitive, *we have always been driven by the compulsion "to know."* "There are no peoples however primitive," Polish cultural anthropologist Bronislaw Malinowski wrote in the opening lines of a 1925 article on magic, science, and religion [2, p. 1], "lacking either in the scientific attitude or in science...." Survival in a "primitive community," he continued, would not be possible "without the careful observation of natural process and a firm belief in its regularity, without the power of reasoning and without confidence in the power of reason; that is, without the rudiments of science." It seems that science, a word derived from the Latin, *scientia*, meaning "knowledge," is in our genes.

These early inklings, rich in cultural diversity, later matured into grand mythic narratives, many of which concerned ultimate origins—ours and the world we are a part of—and many of these "creation myths" looked to the heavens [3]. Distinguished from widely held but false beliefs or ideas, cultural *myths* are essentially a form of ritualized story telling (Greek, *mythos*: story), narrating the strange in terms of the familiar for the purpose of organizing human experience and behavior. Regarded as conveying profound truths either metaphorically, symbolically, or literally, they are a cultural response to a universal human need to understand ourselves and the world around us in order to feel more "at home in the Universe" (to borrow from the title of Stuart Kauffman's 1995 book), often invoking occult powers or agents, anthropomorphized forces in nature, or the plans of a Supreme Being. Just as the authors of Genesis attributed the occurrence of the Flood to a wrathful God, Neolithic (New Stone Age) societies, dominated by the vagaries of agrarian cycles where the course of events was clearly determined by remote agencies, must certainly have imagined supernatural/superhuman forces directing the natural forces in the world around them.

Astrology, the study of celestial events, typically some combination of solar, lunar, planetary, or stellar alignments, as portents of earthly matters— "astronomy brought down to Earth and applied to the affairs of men," as the nineteenth-century American essayist Ralph Waldo Emerson reputedly remarked—is a type of "sky myth" that traces its beginnings to the ancient practice of divination [4]. This occult pseudoscience—"this ... excellent foppery of the world," declared Edmond in Shakespeare's *King Lear*, "that when we are sick in fortune ... we make guilty of our disasters the Sun, the Moon, and stars"—was inseparable from the science of astronomy until the seventeenth century, and therefore had practical and intellectual dimensions as well (recall Note 4 of Chap. 1, and note that the word "disaster," meaning "bad star," has an explicit astrological connotation). Faith in the rule of celestial bodies over our lives was not unreasonable (Fig. 2.1; Shakespeare's Romeo and Juliet as a pair of "star-cross'd lovers" is another example from his repertoire), given the cyclical governance evident in all aspects of the seasonal year. Indistinguishable until early modern times from the science of astronomy, the oldest of the sciences, astrology is recognized as the seat of science itself.

Creation myths—symbolic narratives of how the habitable world began— are found throughout human culture past and present. They are the most common form of myth and they come in many flavors. The Genesis creation narrative (Fig. 2.2; the word "genesis" is from the Greek meaning "creation, beginning, origin") is central to the Judaic-Christian-Islamic tradition, and thus remains a central creation myth today. Here the concern is not with the origin of matter itself, but rather with its arrangement—creating order (*kosmos*) from disorder (*chaos*)[4]—and function. A similar scenario unfolds in several other creation myths from around the world, including the Hindu account known as the *Purusha Sukta*, when the god Brahma, the "Creator," emerges from a preexisting state of chaos and darkness without form or structure and creates the Universe by ordering the elements and fashioning the first living beings. Similarly, the *Enuma Elish* is an ancient Babylonian creation

---

[4] Our word "gas," a collection of freely flying atoms or molecules moving in a disorderly, chaotic fashion, shares the same root as "chaos." It should be pointed out that, according to the Second Law of Thermodynamics, which states that *entropy*, a measure of disorder, always increases in any closed system (such as the Universe), the natural direction of change is from order to disorder, thus rendering the Creation highly improbable [see, e.g., 5]. While we may indeed be children of the stars, "We are" also, author Peter Atkins reminds us, "the children of chaos, and the deep structure of change is decay" [6, p. 200].

Fig. 2.1 Zodiac man. The connection between the human microcosm and the celestial macrocosm—the belief that man is the microcosmic reflection of the order in the macrocosmic Universe as a whole—is explicit in this beautiful illustration from the famous *Très riche heures du duc de Berry*, an early-fifteenth-century "book of hours" (ca. 1416) now in the Musée Condé, Chantilly, France. Each one of the 12 zodiacal constellations is depicted next to the body part it was thought to control, running down from the head (Aries) to the feet (Pisces) in the same order as they appear in the sky along the zodiac. The Latin inscriptions in the four corners expand on the medicinal properties of the zodiacal signs. By the time of the Renaissance, astrological medicine—"practical astrology"—rooted in antiquity, had been developed to a high degree, as can be inferred from the motto of the School of Medicine at the University of Bologna: "A doctor without astrology is like an eye that cannot see." As Francis Bacon, the Renaissance prophet and promoter of the New Science, wrote: "No natural phenomenon can be adequately studied in itself alone—but, to be understood, it must be considered as it stands connected with all Nature." (*Wikimedia Commons, public domain*)

myth, one of the earliest known, about how the world and humans came into being from a primordial state of chaos personified by the mother goddess Tiamat who is defeated by Marduk (Zeus and Jupiter in Greek and Roman mythology), the chief god of the city of Babylon. Marduk tears Tiamat in half, and from the two pieces he creates heaven and Earth, bringing order to the Cosmos, setting the paths of the planets and stars, and, with Ea, the god of water and wisdom, he creates humankind. In Greek mythology, the creation of the Universe is attributed to the primordial deities, Chaos, Gaia, and Eros. Gaia, the personification of Earth, emerged from Chaos, an initial state of void and darkness; Eros represented the force of love and attraction. In Chinese creation myth, creation occurs as a result of the interaction of opposites, a common theme in Chinese philosophy (Fig. 2.3).

The cosmic mythic image is reflected on a grand scale in the architecture of past and present civilizations. The deliberate astronomical orientations of buildings and ceremonial centers (such as Egypt's Great Pyramid) and sometimes entire cities (for example, Washington, D.C., conceived as the quintessential heavenly city in the post-Enlightenment era), most notably

a   b

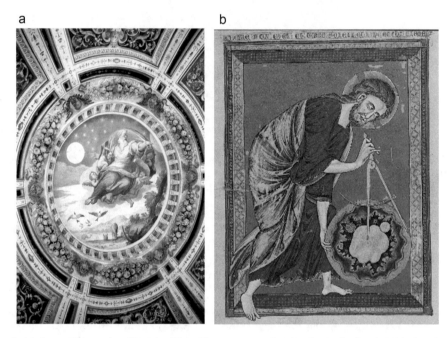

Fig. 2.2 **a** Frescoed ceiling in the Villa Farnese, Caprarola, Italy (ca. 1560), by the Italian painter Taddeo Zuccari (1529–1566) depicting the Genesis account of Creation, similar to Michelangelo's more famous representation on the ceiling of Rome's Sistine Chapel. (*Photograph by the author*), **b** Frontispiece of an Old French *Bible moralisée* (ca. 1208–15; attributed to the atelier of the Catalan Master of St Mark), among the most expensive manuscripts ever produced in medieval Europe, now in the Austrian National Library, showing God as the supreme architect of the Universe. Creation is accomplished with the assistance of a mason's compass/divider: "…and in his hand / He took the golden compasses, prepared / In God's eternal store, to circumscribe / This Universe, and all created things" (Milton, *Paradise Lost*, vii. 224–227). Recall, also, Proverbs 8:27—"he set a compass in the face of the deep." The scene echoes the creation of the world in Plato's *Timeaus* in which a Divine Craftsman impresses geometric forms on preexisting unformed matter. The text above the image reads "ICI CRIE DEX CIEL ET TERRE, SOLEIL ET LUNE ET TOZ ELEMNTZ" ("Here God creates the Heavens and Earth, the Sun and the Moon, and all the Elements"). (*Wikimedia Commons, public domain*)

Fig. 2.3 A mid-eighth-century Tang dynasty depiction of the half-human, half-snake Chinese creation gods Nüwa and her brother-consort Fuxi depicted against a background of constellations with the Sun overhead and the Moon below, now in China's Xinjiang Museum. Creation occurs as a result of the interaction of opposites (here, the male and female principles shown intertwined in double-helix fashion), a common theme in Chinese philosophy. (*Wikimedia Commons, public domain*)

with respect to the north, south, east, and west cardinal directions—the word "orient" means to arrange with reference to the rising Sun in the east—reflects our innate desire to be connected to the Cosmos, evidence that the primitive mythic image is still with us in modern times [e.g., Ref. 3]. Our true cosmic connections will be revealed later in this and the following chapters where we explore the discoveries of modern "intellectual" astronomy.

The ancient Greeks were the first to believe that the world is *rational*—that it is intelligible and understandable in terms of universal, naturalistic explanations. The Greeks invented *philosophy*—literally, "love of wisdom," the systematic study of the nature of knowledge, reality, and existence. Greek thought, the triumph of *logos* (logical reasoning) over *mythos*, was strikingly secular and left no room for capricious and anthropomorphic gods.

This dawn of rational speculation, which accompanied the banishment of the gods to Olympus, gradually dispelled the mist of myth and superstition that had enveloped earlier traditions that typically viewed nature and natural processes as the expression and action of whimsical deities. There were invariant patterns in the workings of nature governed by impersonal, *natural* agents—"laws of nature"—and the Greeks believed that, mortal as we are, we *can* figure them out. *This conviction, indeed faith, in the existence of an* ordo rerum *("order of things")—a* cursus communis naturae *("common course of nature")—the confidence of order and stability within a changeable world—and trust that the ultimate nature of things excludes the supernatural and the arbitrary, together with the confidence that the world* is *rationally intelligible—that the Universe is comprehensible, and that we are not at the mercy of mystical forces or capricious deities; that there is a resonance between the way we think and the way the world works—is perhaps the most important and most profoundly liberating idea in human experience; indeed, this is the real lesson of science.*[5]

Greek rationalism—"the seed of human faith in a Universe that was lawful and regular rather than the plaything of temperamental deities bent upon vengeance and never-ending internecine battles," as author John Barrow put it [8, p. 11]—is rooted in the regularity and predictability of the heavens: the regular motions of the stars are among the most common loci for laws of nature and were regularly taken as the very paradigm of ordered, lawful behavior in antiquity. In addition to offering important practical benefits, such as keeping time (calendrics) and knowing place (navigation), regularities in the heavens promoted psychological influences about the regularity and reliability of nature.

The ancient Greeks were the first to demand that *cosmology*, originally the study of (*-ology*) order in the world (the Greek word *kosmos* originally meant both world and order, since they believed the world was an ordered one), should look to the sky and explain the details of celestial phenomena. Today, cosmology is an important branch of astronomy that investigates the large-scale structure, origin, and evolution of the Universe; the study of the origin

---

[5] Writing about "the Ideas that Have Shaped Our World View," to steal the subtitle of his book [7, p. 47], Richard Tarnas concludes that "The belief that the Universe possesses and is governed according to a comprehensive regulating intelligence, and that this same intelligence is reflected in the human mind, rendering it capable of knowing the cosmic order, was one of the most characteristic and recurring principles in the central tradition of Hellenic thought." (Of all the books I've read—and I've read many—I learned more from this one book than any other—and I thought I already knew a lot about "the Western Mind." It's the best intellectual history of the West I've ever read. Give it a read: you'll learn a lot.)

of astronomical bodies, such as planets and stars, and systems, including the Universe itself, is properly called *cosmogony* (Greek, *-gonia*: begetting). (The late English theoretical physicist and cosmologist Stephen Hawking, stating that his "goal is simple," sought "a complete understanding of the Universe, why it is as it is and why it exists at all," hardly a "simple" goal for most mortals.) In his book *The Copernican Revolution*, a study of the role of planetary astronomy in the development of Western thought, science historian Thomas Kuhn remarked that the Greek

> requirement that a cosmology supply *both* a psychologically satisfying worldview *and* an explanation of observed phenomena ... has channeled the universal compulsion for at-homeness in the Universe into an unprecedented drive for the discovery of scientific explanations [9, p. 6].

Kuhn continues:

> The tradition that detailed astronomical observations supply the principal clues for cosmological thought is, in its essentials, native to Western civilization. It seems to be one of the most significant and characteristic novelties that we inherit from the civilization of ancient Greece [9, p. 27].

The Greek insistence on astronomical cosmologies established a priority for the astronomical sciences that reaches all the way to the early modern period and continues even today. To the extent that the heavens are associated with divinity, cosmology has ever been entangled with *theology*—the study of the nature of God and religious belief—oftentimes to the detriment of each.

The greatest philosopher of ancient Greece was Plato (427–348/7 BC; Fig. 2.4). Indeed, it has been said that all of Western philosophy is nothing more than a footnote to Plato. Plato's cosmology, outlined in his creation narrative, *Timaeus*, posits a single, unique, animate world, created, organized, governed, and maintained by a Divine Craftsman, the Demiurge (literally, "worker of the people"), a thoroughly rational and hence good intelligence endowed with reason and soul (Latin, *anima*; hence *anima mundi*, "World Soul," the inner vitality and life force of the world),[6] as the mainspring and immortal governing principles. The Demiurge creates the world from coexisting unformed matter, first fashioning matter into regular triangles

---

[6] A modern version of Plato's "World Soul," a world fully alive at every level, is the Gaia hypothesis proposed in the 1970s by the British scientist James Lovelock, which argues that Earth (Greek, *Gaia*) is not simply an environment *for* life, but rather *is* life itself, a global living organism and self-sustaining system that modifies its surroundings in ways that ensure its survival, a process that may operate throughout the Universe and is therefore of interest in the search for extraterrestrial life (see Chap. 4).

(mathematical "elements") that combine into the five regular polyhedra—the tetrahedron (a pyramid formed from four equilateral triangles), the hexahedron (a cube, each side of which is made from two isosceles triangles), the octahedron (eight equilateral triangles), the dodecahedron (twelve equilateral pentagons, each made from triangles), and the icosahedron (twenty equilateral triangles)—to give the Empedoclean material elements that finally form the diversity of more complex mixed substances.

Empedocles of Agrigento (fl. 450 BC), a Greek colony in Sicily, argued that the diversity of phenomena in the world required four primary substances—elemental earth, water, air, and fire. To these four "terrestrial" elements found only in the earthly realm, Plato's student, Aristotle (384–322 BC; Fig. 2.4), the most influential natural philosopher of antiquity (Dante called him "the Master of those who know") whose ideas dominated all of Western science until the early modern period in the seventeenth century, added a fifth: a pure, immutable, "celestial" element identified as the *ether* or

Fig. 2.4 Raphael's frescoed *School of Athens* (1510–11) in the Vatican's Stanza della Segnatura, showing Plato with his cosmological treatise *Timaeus* in hand, pointing to the heavens, the realm of Ideal Forms. To his left is his student Aristotle, motioning downward to the reality of the material world where his science of empirical observation must begin. Notice that whereas down-to-earth Aristotle, wearing brown and blue clothing representing the base elements earth and water to signify that his philosophy is grounded and material, has his two feet planted firmly on the ground, Plato, robed in red and grey, representing the imponderable higher elements fire and air, lifts his feet upward in tiptoe fashion. (For more information on this most exquisite work of art, see Christiane Joost-Gaugier's "Plato and Aristotle and Their Retinue: Meaning in Raphael's *School of Athens*," *Gazette des Beaux-Arts*, cxxxvii, pp. 149–164, 2001.) (*Wikimedia Commons, public domain*)

Fig. 2.5 The four elements of the ancient Greeks—earth, water, air, and fire—are depicted in this mosaic made by artist Walter Eglin between 1938 and 1946 at the Kollegienhaus der Universitat of the University of Basel in Switzerland. The crescent Moon and star in the upper left represent Aristotle's fifth element, the quintessence, found only in the celestial realm. (*Photograph by the author*)

quintessence, the latter from the Latin *quinta essentia* ("fifth"—and ultimate—essence; the word is used today to denote the most perfect manifestation of a quality or thing; recall, for example, the 1997 Bruce Willis film, *The Fifth Element*; Fig. 2.5).[7]

According to Empedocles, bone, for example, is a particular mixture of earth, water, and fire, whereas burning wood, composed of ash (earth), tar (water), smoke (air), and flame (fire), is a mix of all four "roots" as he called them, emphasizing their assumed animate character. Sensible change was the result of a change in the proportion of "The Four"—often considered in very general terms with water being the fluid principle and earth the principle of solidity—just as different foods result from different mixes of flour, sugar, water, and eggs.

Although Plato's Divine Craftsman, a transcendent intelligence that rules and orders all things, was neither omnipotent nor providential, Christians

---

[7] Interestingly, the four terrestrial elements are analogous to the four states of matter recognized today: earth (Plato's cube) = solid; water (Plato's icosahedron) = liquid; air (Plato's octahedron) = gas; and fire (Plato's tetrahedron) = plasma, a hot, energetic gas. The quintessence (Plato's dodecahedron) finds its modern counterpart in the "dark" matter-energy that accounts for nearly 95% of the Universe. In more modern times, the ether was resurrected as a pervasive and imponderable medium whose vibrations were thought responsible for various electromagnetic effects including light, only to be dismissed by the genius of Albert Einstein in the early twentieth century (even if we've retained the pervasive term "ethernet" for our modern medium of electronic communication). As we understand physics today, *all the stuff we see*—normal baryonic matter—is made up of just three elementary particles: the "up" quark (U) and the "down" quark (D), subatomic particles that combine in triplet to form protons (UUD) and neutrons (DDU), and the electron, fundamentally simple—and therefore something the ancient pre-Socratic Greek philosophers surely would have appreciated.

would later welcome the *Timaeus*, Plato's only work known to the Latin Middle Ages, because of its resounding resonance with revealed theology: the world is created (not eternal) through intelligent design (not by chance) by a single god (not a pantheon), and that god is eternal, good, and pleased with the Creation.

Anaxagoras, who introduced philosophy to Athens, was the most important astronomer of the fifth century. Using only basic geometry, he gave the first correct explanation for eclipses, both solar and lunar, and, arguing correctly that the Moon shines by reflected sunlight (becoming Shakespeare's "errant thief… snatch[ing light] from the Sun"), he explained how the Moon goes through its cycle of phases. He reasoned further that because the light from the Sun is not reflected mirror-like from the Moon (no image of the Sun is produced), but is instead reflected diffusely as from a rough, irregular surface, the Moon's surface must be dirty like Earth's. In this rather simple but powerful application of principles of physics—here, the laws of geometric optics—to an astronomical object and environment, Anaxagoras became history's first astrophysicist. Two thousand years later, Galileo's telescope demonstrated observationally the Moon's likeness to Earth, and only in 1969 did we "do the experiment" and travel to the Moon to find that its surface is indeed very dusty. Extrapolating this important finding beyond the Moon to other astronomical bodies suggested *a universality of "stuff" within the Cosmos*, which, in turn, suggests that *a uniform and universal set of laws might govern both terrestrial and celestial phenomena*. Both of these interpretations—the universality of matter and of laws—correct we now know, contradict, as we shall see, Aristotle's separation of the terrestrial realm from the celestial.

Anaxagoras identified the "first principle of movement" in the Cosmos with *Mind* or *Intellect* (Greek, *Nous*; $\nu o \upsilon \varsigma$), the directing intelligence of the Cosmos, which in time became Plato's "World Soul," Aristotle's "Prime Mover," and the angels and assorted "intelligences" of the Middle Ages. The presence of this rational soul in nature, the macrocosmic equivalent of the microcosmic human soul, was central to the Greek view of the Universe as an intelligent, living organism, and was regarded as *the source of regularity and orderliness in the natural world that makes a science of nature possible. The lawfulness exhibited by the heavens was believed to be but one manifestation of a pervasive underlying lawfulness operating throughout all of nature*.

This faith that there is order and reason at the heart of things continues to be a vital force in the scientific thought of the Western tradition.

It should be no surprise that the Greek Cosmos was a geometric Cosmos, built upon the axioms of their own invention. One of Plato's students,

Eudoxus of Cnidus (fl. 375 BC), was the most brilliant mathematician of the fourth century, indeed one of the greatest of the classical period (his work forms the essence of several parts of Euclid's *Elements*, a great compendium of geometry, still studied in schools today). He was the first to construct a mathematically-based (geometric) model of the Universe that required a total of 27 onion-skin-like nested concentric and *geocentric*—Earth-centered— crystal spheres to reproduce the combined diurnal (daily) and annual (yearly) motions of the planets and stars. Although there were problems with his model—it could not account for the observed change in the angular size of the Moon, nor could it explain the increased brightness exhibited by the planets during their retrograde (eastern) motion—it certainly demonstrated to posterity the power of geometric models in science.

Aristotle refined the system of Eudoxus by including additional spheres to "unroll" the motion of the other planets, requiring 55 spheres in all, including the sphere of Earth. Aristotle's geocentric Cosmos, centered on Hamlet's "goodly frame, the Earth," remained popular for two thousand years: Fig. 2.6 is taken from the leading astronomy book of the early sixteenth century, Peter Apian's *Cosmographicus liber*, and is essentially Aristotle's geocentric Cosmos (here showing only the primary spheres).

Aristotle's corporeal Universe was of necessity finite in spatial extent, quite small and compact. Otherwise, the daily motion of the spheres would proceed with infinite speed. Furthermore, an infinite Universe would have no center at which to place Earth. Aristotle's Universe was limited in space but, unlike Plato's created Universe, unlimited in time "for the Sun and stars are born not, neither do they decay, but are eternal and divine." For Aristotle, the world must be eternal because the alternative—creation *ex nihilo* (out of nothing)— was untenable.

With the founding of Alexandria's great Library and Museum in the early third century BC, the intellectual center of the Western world shifted from Athens to Alexandria, where it remained for nearly a thousand years until it was eclipsed by the great Islamic centers of learning. It was here that the "Copernicus of antiquity," Aristarchus of Samos (ca. 310–230 BC), suggested a *heliocentric* (Sun-centered) model of the Universe, which was, however, rejected and forgotten for nearly two millennia on the basis that if Earth moves around the Sun, the relative positions of the stars on the sky should shift when they are observed at different times of the year and hence from different points in Earth's orbit, with closer stars exhibiting greater variation just as you would notice nearby objects appear to shift back and forth when swinging your head from side to side. No such changes, known as *stellar parallax*, were observed in antiquity because the greatest shift, a minuscule

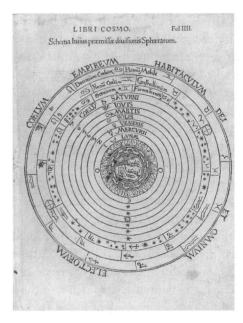

Fig. 2.6 An illustration of the geocentric Universe taken from the leading astronomy book of the early sixteenth century, Peter Apian's 1533 *Cosmographicus liber*, a work that went through no fewer than thirty editions in fourteen languages in that century alone. Note the location of the four terrestrial elements, including Hamlet's "most excellent canopy, the air," beneath the orb of the Moon, beyond which lies Shakespeare's "brave o'erhanging firmament, this majestical roof fretted with golden fire." (*Wikimedia Commons, public domain*)

1½ arcseconds (about the angular size of a penny viewed from a distance of a mile and a half) for the nearest star system, Alpha Centauri, was far too small to be detected by naked-eye observations which are limited to about 10 arcminutes, about one-third the angular size of a full Moon. The stars were simply too far away, much father than the ancients had imagined. (The first detection of stellar parallax, possible only with a telescope, was not reported until 1838, and the technique is still used today to determine stellar distances.) Even more damaging than the *observationally* based arguments, heliocentrism was inconsistent with Aristotle's *theory* of physics which was predicated on the teleological (goal-directed) notion of "natural" places and motions. Of the four terrestrial elements, earth was endowed with absolute heaviness and hence should seek its proper station at the center of the Universe. Astronomy was clearly subordinate to physics. Finally, a moving Earth violates *values*—common sense and authority—Aristotle's and Plato's, both of whom believed the planets were divine beings and that heaven is the visible image of the divine, and thus, while open to rational speculation, the province of the theologian. Although heliocentrism was rejected in antiquity, it is noteworthy that it was rejected on what were, at the time, sound scientific arguments concerning observation and theory, the essence of science even today. Besides, Earth feels fixed, permanent, immobile, with the entire Cosmos revolving around us. It's just "common sense."

Four centuries later, based on earlier work by Apollonius of Perga (fl. 210 BC), "The Great Geometer," and by Hipparchus of Nicaea (ca. 190–ca. 120 BC), proclaimed to be the greatest observational astronomer of antiquity, the Alexandrian sage Claudius Ptolemy (ca. AD 90–170) proposed a modified geocentric cosmology that incorporated various devices—epicycles, deferents, eccentrics, and the equant (Fig. 2.7)—to improve predictions of planetary motion, mostly for the purpose of making more accurate astrological predictions. Ptolemy was the greatest astronomer of antiquity—his 13-volume cosmological compendium has come down to us through medieval Arabic translations as the *Almagest* (Arabic, *The Greatest*)—and the greatest astrologer: his astrological treatise *Tetrabiblos* (Greek, "four books") was literally the "bible" of astrology.

Supported by the authority of Aristotle and later by the Catholic Church (in the thirteenth century, soon-to-be-saint Thomas Aquinas pronounced that speculation on whether "the world had a beginning ... is an article of faith, not of demonstration or science"), Ptolemy's menagerie of epicycles,

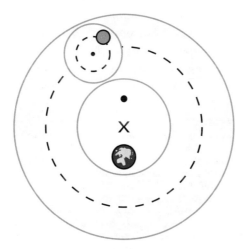

Fig. 2.7 The basic elements of Ptolemaic cosmology showing an orange planet on an epicycle (smaller dashed circle), the center of which revolves on a deferent (larger dashed circle) around the center (×). The equant (•) is an off-centered point across from Earth which is located at the eccentric point an equal distance from (×) as is the equant, about which the motion of a planet proceeds at a constant ("equal") angular speed. (*Wikimedia Commons, public domain*)

deferents, eccentrics, and equants, together with Aristotle's model of concentric spheres, dominated astronomical thought throughout the Middle Ages and the Renaissance. Each represented the two opposing views of nature dating back to the beginning of Greek science (recall Fig. 2.4), with the realistic material/physical Cosmos of Aristotle at one pole and the ideal/ mathematical Ptolemaic astronomy at the other, each characteristic of two distinct ancient civilizations: the earlier, Hellenic and physical; the later, Hellenistic and mathematical. Ptolemy's model provides an *instrumentalist* view—an unphysical but useful and convenient fiction devised expressly as a computational device "to save the phenomena," much like most navigational calculations today that are typically referenced for convenience to a fictional stationary Earth. Aiming only for empirical adequacy, instrumentalism, a form of antirealism strongly opposed to philosophical realism, "regards scientific theories as instruments for helping us predict observable phenomena, rather than as attempts to describe the underlying nature of reality," according to philosopher of science Samir Okasha [10, p. 56]. Until the early modern period with the likes of Kepler, Galileo, and Newton, astronomers were mere mathematicians not engaged in explanatory natural philosophical matters. Aristotle's Cosmos wasn't much better: a goal-directed, compact, geocentric Universe made of just five elements: earth, water, air, fire, and the celestial quintessence (recall Fig. 2.5).

While perhaps seeming highly contrived to the modern mind—even the thirteenth-century Castilian king, Alfonso X, was said to have remarked, in an appeal to the principle of parsimony, that if the Lord had consulted him before embarking upon Creation, he would have suggested something *simpler!*—to the classical mind, the simplest motion was unchanging, uniform circular motion.[8] Dismissing as guesswork any speculation about

---

[8] The notion that "simpler is better" was formalized by the fourteenth-century Oxford scholar, William of Ockham, and is known in his honor as *Ockham's razor* (to cut away the superfluous from the essential), a precept embraced, incidentally, in the modern world of industrial design by Apple's Steve Jobs whose Zen-like minimalist mantra was "Let's make it simple. Really simple." This "economy of nature"—the belief that God or nature (many would make them one and the same) operates with the fewest possible causes—is reflected in an economy of thought that can be identified as a *principle of parsimony*. "*Natura nihil facit frustra*," Newton wrote (and that's in Latin so it must be true!): "Nature does nothing in vain," a perception shared by so many others throughout the history of science. In Aristotle's time, Heraclides of Pontus (ca. 388–310 BC), Plato's successor at the Academy, simply suggested that it would be a lot simpler, in the sense of being easier and more economical, to allow Earth to rotate once per day rather than to move all those celestial spheres. (Heraclides's suggestion also raises a *caution with regards to sensory perception*: if Earth rotates, it is only an *apparent* motion of the heavens that is perceived on Earth.) A rotating Earth also admits an infinite Universe: if the heavens aren't rotating, they could extend, in principle, to infinity, thus relieving Aristotle's concern over the infinitely fast rate of rotation of infinitely big spheres. Lots of far-reaching consequences for such a relatively simple suggestion, and yet it was not generally accepted because, it was argued, such a rapid rotation of Earth would produce unobserved marked

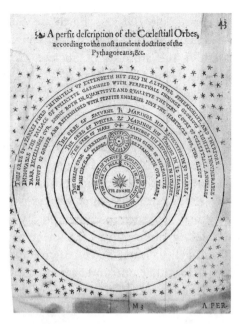

Fig. 2.8 The infinite Copernican Universe of Thomas Digges, reproduced from his 1576 *A Perfit Description of the Caelestiall Orbes*. Digges shared the same London neighborhood as Shakespeare, whose Hamlet speaks of being "a king of infinite space." Close inspection of the sublunary sphere surrounding Earth shows that the new astronomy is accompanied by the old physics: Aristotle's hierarchical arrangement of flames of fire reach down from the orb of the Moon to clouds of air floating above the "globe of mortalitye." (*Wellcome Collection, attribution 4.0 International CC BY 4.0*)

the underlying fabric of the heavens, Ptolemy was committed to "saving the appearances" by mathematical certainty and simplicity, and in a show of intellectual modesty, he believed that the power of the human mind was limited, and that a true understanding of the structure of the heavens belonged only to divine intelligences.

More than a millennium later, the Polish astronomer Nicolaus Copernicus (1473–1543) resurrected the heliocentric cosmological *hypothesis* (Fig. 2.8)—although the number of sixteenth-century Copernicans could be counted on the fingers of one's hands. Significantly, as the seventeenth-century Anglican clergyman and natural philosopher John Wilkins asked in his 1640 *Discourse*

---

effects—such as suggested by rocker Neil Young in his song, "Comes a Time": "Oh-oh, this ol' world keeps spinnin' round / It's a wonder tall trees ain't a laying down." In more modern times, Einstein believed that, apart from any mathematical formalism, a good theory in science should be simple enough to be understood by a child; for the New Zealand-born British physicist Ernest (Lord) Rutherford, it had to be simple enough for a barmaid.

Concerning a New Planet, "Now if our Earth be a Plannet, then why not another of the Plannets be an Earth?", thus reviving the plurality of worlds debate initiated by the atomists of ancient Greece. Based on the *observations* made by the Danish astronomer Tycho Brahe (1546–1601), the most accurate pre-telescopic observations of the heavens ever made, the German astronomer Johannes Kepler (1574–1630) discovered *empirical laws* governing planetary motion, finding that the planets move in elliptical orbits at variable speed, thus finally ridding the stables of astronomy of the Platonic perfection of uniform circular motion. And finally, "standing on ye shoulders of [these] Giants" and others before him, including the Italian astronomer Galileo Galilei (1564–1642; born the same year as William Shakespeare, to relate all of these events in science to contemporary humanistic achievements), the English "natural philosopher" (as scientists were called before the mid-nineteenth century) Sir Isaac Newton synthesized all of this into a *theory* to explain it all. Note that our understanding of planetary motion, the problem that lay at the heart of the so-called Scientific Revolution that lifted us into the modern world [11], unfolding here over the course of nearly a century and a half, traces a particular scientific method progressing from hypothesis → observation/experimentation ("hypothesis testing") → empirical law → theory.

And so, knowing how we got here, we now turn to modern developments in cosmology.

2.2   Modern Cosmology

The point of view of a sinner is that the church promises him hell in the future, but cosmology proves that the glowing hell was in the past.
— Soviet physicist Ya. B. Zel'dovich (1914–1987; quoted in J. Silk, *The Big Bang*, p. 109)

The Universe consists of 5% protons, 5% neutrons, 5% electrons, and 85% morons.
— American musician Frank Zappa

In ancient Greece, Plato taught that the world was created in time; his pupil Aristotle taught that it was eternal and unchanging and thus timeless. Twenty-four centuries later, the nature of the Universe and its origin—and by implication, *our* ultimate origin, purpose, and destiny, certainly *the* most fundamental of questions forever on the minds of humankind (recall

Fig. 1.1)—were still being debated, now for the first time scientifically by modern cosmologists, some of whom, like Plato, argued for a big-bang beginning, while others favored, like Aristotle, an unchanging steady-state Universe that always was and always will be. Thanks largely to the work of the American astronomer Edwin Hubble (1889–1953; Fig. 2.9), who discovered the true "realm of the nebulae" (to borrow from the title of his popular 1936 account of his discoveries; the singular form "nebula," is from the Latin meaning "mist" or "cloud"), we now know that the Universe—space and time itself—began with a Big Bang[9] some 13.8 billion years ago. During the first three minutes of its history, all of the hydrogen and most of the helium—by far the most abundant elements in the Universe which is roughly three-quarters hydrogen and one-quarter helium by mass, an abundance ratio determined by the slight difference in the mass of the proton and the neutron—formed from a cooling plasma of protons, neutrons, and electrons (the morons appear much later).[10] We now know that stars and

---

[9] Actually, the term "Big Bang" is a misnomer: spacetime was non-existent, indeed brought into existence, with the creation of the Universe, so there was no medium through which sound could travel. It all began, it seems, much more like a "Quiet Whisper," a much more appropriate, if less powerful, label for creation.

[10] *Protons* (*p*), positively charged subatomic particles, and electrically neutral particles called *neutrons* (*n*), which are only slightly more massive than protons, are called *nucleons* because they are the constituents of atomic nuclei. Although stable when bound in an atomic nucleus, a free neutron is unstable to *beta decay* (β-decay), a type of radioactive decay with a half-life of about 15 min during which the neutron is transformed into a proton with the emission of an *electron* (*e*) and an antineutrino: $n \rightarrow p + e + \bar{\nu}$, where the bar denotes the antiparticle, in this case the (electron) antineutrino (the *neutrino* is a nearly massless subatomic weakly interacting electrically neutral particle represented by Greek letter ν, "nu"; its name is from the Italian for "little neutral one"). The decay of carbon-14 (^{14}C) into nitrogen-14 (^{14}N) with a half-life of about 5,730 years is a familiar example of β-decay used to date archeological artifacts. Protons and neutrons are made from small, more fundamental particles called *quarks*, which come in six "flavors," only two of which, the "up" quark (U) and the "down" quark (D), combine in triplets to make up the proton (UUD) and the neutron (DDU; recall Note 7). Negatively charged electrons, nearly 2,000 times lighter than nucleons and exhibiting no internal structure, are found in the outer parts of atoms and are the primary carriers of electric current. The positively charged *positron* (e^+) is the electron antiparticle; antiparticles have the same mass but opposite electric charge of their paired (charged) particle.

The electron, the first elementary particle to be discovered, was first detected by the English physicist J. J. Thomson (1856–1940) in 1897; the proton in 1919 by the New Zealand-born physicist Ernest Rutherford (1871–1937) who also discovered the nucleus of the atom in 1911. Both the neutron and the positron were discovered in 1932; the former by the English physicist James Chadwick (1891–1974), the latter by the American physicist Carl Anderson (1905–1991). The elusive neutrino was finally detected in 1956 by the American physicists Clyde Cowan (1919–1974) and Frederick Reines (1918–1998) after being predicted by the Austrian physicist Wolfgang Pauli (1900–1958) in 1930. All of these men—and they were almost always men back then—were awarded the Nobel Prize, all in physics except for Rutherford's in chemistry for his "investigations into the disintegration of elements and the chemistry of radioactive substances" (Cowan was deceased and so shared it with Reines in name only). Rutherford's chemistry Nobel Prize in 1908 came with a certain engaging irony: he reportedly once quipped that chemistry is what stinks and that chemists and damned fools were one and the same! During the award ceremony, he joked that he had seen many transformations in his work, but the quickest was his own from a physicist to a chemist. Today,

galaxies formed relatively early in the history of the Universe, aided by a mysterious, nonluminous substance we call "dark matter" that shows itself through gravity, not light (recall Note 5 of Chap. 1). The first generation of stars began the transition from primordial Big Bang *nucleosynthesis*—literally, the synthesis of the nuclei (of the chemical elements)—to stellar nucleosynthesis, without which we wouldn't be here to write, read, or think about it—or about anything else.

Even the very nature of our own sidereal system and its relation to the Universe-at-large were not known until well into the twentieth century, before which time it was generally believed that our Milky Way galaxy included *everything* in the Universe. Two discoveries made in 1912 were crucial in establishing a new and larger picture of the Universe and our place in it, a bold new view of the Cosmos that, in the opinion of science historians Peter Bowler and Iwan Morus, "ranks as a scientific revolution comparable to one of the defining events of the Scientific Revolution itself" [12, p. 277].

Fig. 2.9 "American Scientists" first day of issue U.S. postage stamp honoring the American astronomer Edwin Hubble, the namesake of the orbiting Hubble Space Telescope, with California's 200-inch Palomar telescope in the background. (*Photograph by the author from author's collection*)

---

the situation is reversed, and it is sometimes said—only half jokingly—that chemistry has become a branch of physics, a state of affairs that would certainly have pleased Rutherford who believed that in science there is only physics, all the rest being mere stamp collecting! Physicists are notoriously scornful of scientists from other disciplines, and chemists seem to be an especially attractive target. Nobelist Wolfgang Pauli was staggered with disbelief when his first wife left him (he wasn't a pleasant person) for a chemist, remarking in wonder to a friend: "Had she taken a bullfighter I would have understood, but a *chemist*...."

The main reason for the uncertainties in both the size of our Galaxy and the existence of extragalactic "island universes," as the eighteenth-century German philosopher Immanuel Kant called them—distant aggregations, we now know, of enormous numbers of stars like our own Milky Way—was the lack of reliable methods for determining distances in the Cosmos, always a problem in astronomy which looks out onto a seemingly flat two-dimensional celestial sphere. In 1912, after studying variable stars, stars whose brightness varies with time, in the Small Magellanic Cloud, one of two nebulous patches in the southern sky first reported by the Portuguese explorer Ferdinand Magellan in the early sixteenth century, Henrietta Swan Leavitt (1868–1921), one of the female "computers" at the Harvard College Observatory, discovered a relationship between the average brightness and the period of brightness variability for a particular class of pulsating variable stars called Cepheids (after the prototype δ Cephei; the North Star, Polaris, is the nearest Cepheid variable). Just as a heavier mass suspended from a spring takes longer to oscillate up and down, brighter stars, typically being more massive, take longer to cycle through their radial pulsation. Because Cepheids can be intrinsically a hundred thousand times brighter than the Sun, this *period-luminosity relation* became one of the most powerful means for determining distances far beyond the limited range of trigonometric stellar parallax techniques. Once it was calibrated shortly after its discovery, the observed period of variability gave the intrinsic brightness of a Cepheid, which, when compared to the star's apparent brightness, gave its distance (much as one can judge the distance of a known source of light by its apparent brightness).

While there was no broad consensus on the dimensions and detailed structure of our Galaxy in the early years of the century, it was widely accepted that it was finite and flattened, the latter probably due to rotation, and either ellipsoidal or, like the spiral nebulae, possibly spiral in form (Fig. 2.10).[11]

---

[11] Upon turning the newly-invented telescope to the Milky Way, Galileo immediately noticed "a vast crowd of stars," settling, at last, "all disputes which have vexed philosophers through so many ages." (The proposal that the Milky Way was nothing more than a compaction of faint stars had been suggested in antiquity. A painting by the Venetian Renaissance artist Jacopo Tintoretto, now called *The Origin of the Milky Way* and now in London's National Gallery, depicts stars being flung into the heavens from a lactating Juno, and was commissioned some thirty years prior to the publication of Galileo's telescopic discoveries in his 1610 *Sidereus nuncius*.) Galileo correctly concluded that "the galaxy is, in fact, nothing but a congeries of innumerable stars grouped together in clusters," a conclusion shared by the contemporary English poet John Milton, writing in his epic poem, *Paradise Lost* (vii. 577–78): "A broad and ample road, where dust is Gold, / And pavement Starrs ..." Thus was initiated the science of galactic astronomy.

The discovery of the multitude of stars invisible to the unaided eye questioned the very purpose of the Creation and the importance of humanity within the created Cosmos: did Genesis not proclaim that everything was created for our enjoyment and benefit? Are we not the focus of Creation? If yes, what then was the purpose of stars hidden from our view for so many millennia? Perhaps (gasp!) we aren't so special after all. But wasn't this precisely the lesson of Copernicanism? The "numberless

Puzzled by the fact that globular clusters—tight globular groupings of up to a million stars—crowded into one hemisphere of the sky concentrated dispro-portionately in the direction of the constellation Sagittarius, the American astronomer Harlow Shapley (1885–1972), who had worked out a period-luminosity law for Cepheids in globular clusters, used the 60-inch telescope on Mount Wilson just outside Pasadena, California, at the time the world's largest, to determine their distances and hence their actual distribution in space. Shapley presented his results in a 1918 paper titled "Remarks on the Arrangement of the Sidereal Universe," announcing that the globular clusters outline a stellar system some 300,000 light years in diameter, with the Sun, until then generally believed—in very non-Copernican fashion—to be at the center of the Galaxy, displaced nearly halfway out from its center which he argued marked the center of the Galaxy. Thus was humanity moved yet again from the center of things, this time out to the suburbs of a huge disk-shaped galaxy containing a few hundred billion stars (a number comparable to the number of grains of rice that could be packed into a medieval cathedral).

In 1930 the Swiss-born American astronomer Robert Trumpler (1886–1956) working at Lick Observatory just east of San Jose, California, revealed that Shapley had overestimated the size of our Galaxy and hence the distance of the Sun from its center by a factor of three because he underestimated the absorption of starlight by interstellar dust, which dims the light from a star leading to an overestimate of the star's distance. By the late 1920s the Swedish astronomer Bertil Lindblad (1895–1965) and the Dutch astronomer Jan Oort (1900–1992) showed from the motions of stars that the Galaxy rotates and does so, like the Sun and most non-solid astronomical bodies, differen-tially with different regions taking different amounts of time to complete one revolution. Their results located the dynamical center of rotation, and thus that of the Galaxy itself, which they concluded must have a very large concen-tration of mass, in the same direction determined by Shapley, although, significantly, at a distance only one-half to one-third of Shapley's estimate. By the early 1950s, radio telescopes, working at the long wavelengths that, like terrestrial radar, penetrate the cloudy component of the Galaxy, revealed its spiral-arm structure. Shapley's globular clusters, we now know, are indeed

---

... innumerable stars ... spangling the hemisphere"—Milton, too, was impressed by their number—punched holes through more than the velvety black expanse of space; they perforated the prevailing notion of a special creation for a special planet and its special human inhabitants. Also, the very fact that stars appear as dimensionless points of light even when viewed through the telescope implies that they are at enormous distances, too far away (with the exception of our star, the Sun) to be resolved into globes. Just as the telescope brought the Moon and the planets closer, it pushed the innumerable stars farther out to immeasurable distances, furnishing further evidence for a possibly limitless Copernican Cosmos.

Fig. 2.10 The Milky Way photographed from Wyaralong Dam Recreational Facility, Bromelton, Australia. The term comes from the Latin, *Via Lactae*; *lac* is Latin for milk, whence lactose, the naturally-occurring milk sugar, originally from the Greek, *galaktos*, milk; whence "galaxy." Galaxies do look milky white through a telescope, as do the great spiral galaxy M31 (see Fig. 2.11a) and the Large and Small Magellanic Clouds, irregular dwarf galaxies visible from the southern hemisphere. Because we are located inside our galaxy, a little more than halfway out from the center, we see stars in every direction when looking at the night sky, but we see a lot more—a milky band of stars and dark, interstellar dust—when we look along the plane of our (disk-like) galaxy. (*Photograph by Rob Musson on Unsplash, public domain*)

distributed in a spherical halo around the center of the Galaxy with an increasing concentration towards the galactic nucleus.

In 1924 Hubble, working with the 100-inch Hooker telescope on Mount Wilson, then the world's largest, was able to identify individual Cepheid variables in the Andromeda Nebula (M31; Fig. 2.11), and estimated its distance to be 800,000 light years, clearly outside even Shapley's oversized Galaxy. A later refinement (it turned out that Hubble was using a period-luminosity law that was applicable to a different, dimmer class of Cepheids found in globular clusters) showed that the Andromeda Galaxy, the nearest major galaxy to our own, lies about 2.5 million light years from our Galaxy, a distance roughly equal to 25 times the diameter of the Milky Way.[12] We are indeed a very

---

[12] Distances, like masses, sizes, densities, ages, energies, and so many other properties of astronomical objects, reach far beyond the scale of everyday familiarity, a reach so far they're typically measured

small and insignificant speck inside an equally insignificant, if larger, speck drifting aimlessly through the vastness of space with countless other galaxies, mere molecules in an unimaginably immense cosmic gas. And *for the first time in human history, it was realized that there is more to the Universe than the Milky Way*. As noted at the time by a leading astronomer, this realization "that our galaxy is not unique and central in the Universe certainly ranks with the acceptance of the Copernican system as one of the great advances in cosmological thought."

In the same year that Hubble announced his discovery, the German astronomer Carl Wirtz (1874–1939) and the Swede Knut Lundmark (1889–1958) published the first results indicating that the recessional velocities of the spiral nebulae increase with increasing distance, using apparent size as a distance indicator (assuming smaller implies farther) and the *Doppler effect* to determine velocities (Fig. 2.12).[13] This discovery was a precursor to what has been called "the most important [discovery] ever in the history of cosmology": the expansion of the Universe, the second important astronomical discovery made in 1912.

Working that year with the 24-inch refractor at the Lowell Observatory in Flagstaff, Arizona, the American astronomer Vesto Slipher (1875–1969) measured the first radial velocity of a spiral nebula, that of M31, finding a value of about 300 km/s, some ten times the largest radial velocity then known to astronomy. By the mid-1920s he had obtained spectra and radial velocities for several dozen nebulae, finding a preponderance of redshifts with only two galaxies, including Andromeda, exhibiting blueshifts. Interpreting

---

in terms of how far light travels in a year: a light year, some six trillion miles—far enough! Even though at 2.5 million light years away it is a relatively nearby galaxy, the enormity of its distance can perhaps be more readily appreciated by remembering that the light reaching Earth tonight from the Andromeda Galaxy, traveling at nearly 300,000 km/s (186,000 mi/s)—fast enough!—left it 2.5 million years ago when our distant hominin ancestors, the Australopithecines ("southern apes"), were mucking about on the East African Rift. (To put this distance on earthly scales, it would take almost 25 *trillion* years—nearly 1800 times the age of the Universe—to reach Andromeda traveling in a car at 70 mph.) This *cosmic look-back time* due to the finite speed of light allows us to "see the past" in cosmology in a much more direct sense than is possible in other historical sciences such as geology and biology. It really is amazing that we can see the fuzzy blob of Andromeda, so far away, the most distant thing we can see with the unaided eye. Andromeda, the only galaxy visible to the naked eye from the northern hemisphere, together with the southern hemisphere's Large and Small Magellanic Clouds, are the only galaxies visible to the eye; the rest are too far away to see without a telescope.

[13] Discovered in 1842 by the Austrian physicist Johann Christian Doppler (1803–1853), this motion effect causes waves to be stretched out to longer wavelengths (hence "redshift" for light, but applicable to any type of wave, such as the observed change in pitch of a siren of a passing emergency vehicle) when the source and the observer of the wave recede from each other, and causes waves to be compressed to shorter wavelengths ("blueshift" for light) for motion of approach, with the amount of shift increasing with increasing line-of-sight velocity. It is one of the most powerful diagnostic principles used in physics and astronomy—and in medicine (Doppler ultrasound) and police Doppler radar guns.

a

b

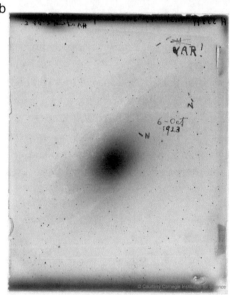

Fig. 2.11 a M31, the Andromeda Galaxy, imaged through the 13-cm telescope within the K'chi Waasa Debaabing Dome at the Killarney Provincial Park Observatory complex, Ontario, Canada. The designation M31 implies that this is the 31st entry in the catalog of just over one hundred nebulous objects including gaseous nebulae (for example, M42, the Orion Nebula; see Fig. 3.5), star clusters (including M45, the Pleiades), and galaxies, compiled in 1781 by the French astronomer Charles Messier (1730–1817; whence "M"). It is the closest major galaxy to our own, about 50% larger in diameter, and it's moving towards our Galaxy at a speed of about 100 km/s leading to an eventual collision between the two in a few billion years. Stay tuned. (*Wikimedia Commons, public domain*), **b** The photographic plate taken the night of 5–6 October 1923 by Hubble at the 100-inch Hooker telescope on Mount Wilson identifying the first Cepheid variable (marked "VAR!", first marked "N" along with two other novae he discovered) in M31 that established the nebula as a galaxy far outside our own Milky Way. (*Photograph courtesy of Carnegie Institution of Science. Used with permission*)

Fig. 2.12 An illustration of the Doppler effect. Waves emitted by a source moving from the right to the left (as indicated by the arrow on the left) are, like a compressed spring, compressed ("blue-shifted" to shorter wavelengths) in front of the source and, like a stretched spring, stretched out ("red-shifted" to longer wavelengths) behind the source. Note that *wavelength* λ is the distance between successive crests (or troughs) of a wave. (*Wikimedia Commons, public domain*)

these observations as Doppler shifts implied that all but two of these nebulae are moving rapidly away from us, many at speeds greater than that required to escape from our Galaxy.

Hubble and his assistant Milton Humason (1891–1972) measured the distance to Slipher's nebulae. From a collection of 18 galaxies plus the Virgo cluster, they discovered—and Hubble announced in a 1929 paper to the National Academy of Sciences—*the single most important fact in modern cosmology*, known as *Hubble's Law*, that a galaxy's distance d is proportional to its redshift which, interpreted as a Doppler shift, is a measure of the galaxy's recessional velocity v:

$$v = H_o d,$$

where the constant of proportionality H_o is the *Hubble constant* (Fig. 2.13a). To many, the cosmological significance was immediate: *our Universe is expanding and therefore could very well have had a beginning*, a discovery that has been called the greatest scientific discovery of the twentieth century. Regardless, Hubble's Law encapsulates *the most important observational fact in modern cosmology—that the Universe is expanding*—and it remains an extremely powerful method for determining remote distances in the Universe: using Doppler's formula to obtain a distant object's recessional velocity from the observed redshift of spectral lines in the light from the object, Hubble's Law gives its distance (Fig. 2.13b).

Don't be deceived into thinking that because most galaxies are flying away from us, that we are (in a very non-Copernican way) at the center of the Universe: receding galaxies are what anyone anywhere in an expanding Universe would see, as blowing up a balloon marked with dots or watching raisins in a rising loaf of raisin bread clearly show (pick any dot or raisin

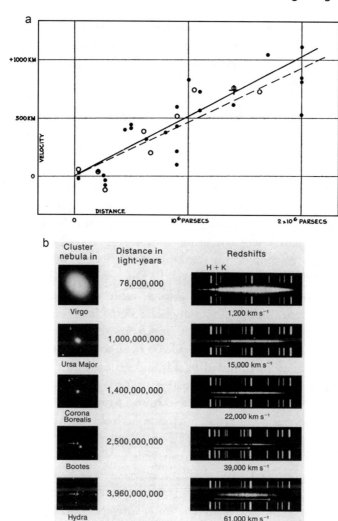

Fig. 2.13 **a** Hubble's original (1929) galaxy velocity–distance relation. Doppler radial velocities (labeled in the wrong units; they should be km/s) are plotted against distances determined from embedded "standard candle" stars and mean luminosities of galaxies in a cluster. The straight lines indicate an approximate linear fit, $v = H_o d$, that became known as Hubble's Law. (*Photograph courtesy of the Edwin Hubble papers, The Huntington Library, San Marino, CA*), **b** Photographs of successively more distant cluster galaxies and their barcode-like spectra (singular, *spectrum*: the variation of light intensity with wavelength), bracketed here with bright spectral lines of iron for reference; atoms of chemical elements emit and absorb light at wavelengths—colors—unique to each element. Note the *increasing* redshift (from top to bottom, with wavelength increasing from left to right, color from blue to red) of the two dark H and K absorption lines due to ionized calcium, and the corresponding *decreasing* size of the source galaxy, both indications of increasing distance (which have since been refined). (*Wikimedia Commons, public domain*)

for your home galaxy and watch all the other "galaxies" move away from yours). And don't worry about an expanding Universe overcoming your reach for your next beer: cosmic expansion, the so-called "Hubble flow," is significant only over cosmological (i.e., HUGE) scales, as the value of Hubble's constant confirms. For the best current estimate of Hubble's constant, roughly 73 km/s/Mpc—meaning that galaxies are receding at a rate of about 73 km/s for each Mpc in distance from us—cosmic expansion amounts to less than one-ten-billionth of a mile per hour over a distance equal to the size of Earth.[14]

Modern theoretical cosmology developed in the wings of Albert Einstein's general theory of relativity (Fig. 2.14), his 1915 theory of gravity that explains the gravitational effect between masses as being due to their warping of spacetime: "matter tells spacetime how to curve, and curved spacetime tells matter how to move," to paraphrase the American theoretical physicist John Archibald Wheeler (1911–2008), he who popularized the term "black hole" for gravitationally collapsed objects (more on these monsters later). In 1917, in a paper titled "Kosmologische Betrachtungen zur allgemeinen Relativitätstheorie" ("Cosmological Considerations on the General Theory of Relativity"), the first application of his new general theory to the entire Universe, work that he admitted to a friend exposed him "to the danger of being confined in a madhouse," Einstein found—or rather forced—a static solution to his field equations. With Hubble's discovery of the expanding Universe still several years away, astronomers told him that the Universe was static. Although his original equations allowed for a dynamic Universe, Einstein introduced a fudge factor, a repulsive force term denoted by the Greek letter lambda (Λ) and called the *cosmological constant*, to counteract the attractive force of gravity that alone would otherwise result in the mutual attraction of everything in a static, finite Universe leading to its ultimate collapse, contrary to its presumed static status. In failing to trust his original equations, with which he could have predicted the expansion of the Universe a decade before its discovery, Einstein missed what would certainly have ranked among the greatest scientific predictions of all time. Einstein later admitted that the introduction of the cosmological constant into his

---

[14] 1 Mpc, pronounced "mega parsec," is a distance of one million parsecs, where one parsec equals 3.26 light years, a HUGE distance when you remember that light travels VERY FAST at a speed of 300,000 km/s (186,000 mi/s) and thus nearly 10 trillion km/yr (6 trillion miles per year), a distance, by definition, of 1 light year. Our Sun is about 8 light-minutes away, since it takes light from the Sun about 8 min to reach Earth, and the nearest star other than our Sun, Proxima Centauri, the closest of three stars in the Alpha Centauri triple-star system, is a little more than 4 light years away. So, the 3.26 million light years in a Mpc, nearly a million times the distance to the next star—and more than thirty times the 100,000 light year diameter of our Galaxy—is indeed a HUGE distance.

field equations was the "biggest blunder" of his life, but in the same breath he reportedly uttered that "death alone saves one from making blunders." Einstein, too, was human after all. But, as we shall soon see, his "blunder" has recently been resurrected to account for the dark energy that makes up nearly three-quarters of the Universe causing its expansion to accelerate. Yes, he was human, but only the blunders of Einstein could have such cosmological importance.

Nevertheless, as science historian Helge Kragh noted in his 1999 history of physics in the twentieth century [13, p. 349], "Einstein's work [is] still considered the foundation of scientific cosmology." And, as it turns out, Einstein's "blunder" is enjoying a renaissance of righteousness, as cosmologists today use it to factor in the accelerating expansion of the Universe, discovered at the very end of the twentieth century—certainly the most consequential discovery in modern cosmology, rated by *Science* magazine the #1 discovery of the year 1998 in all of science and earning its discoverers the 2011 Nobel Prize in Physics—and caused by mysterious *dark energy* which makes up

Fig. 2.14 A cosmological blackboard used by Einstein during a lecture on relativity and cosmology delivered at Oxford University in 1931, preserved today at Oxford's Museum of the History of Science. The first three lines establish an equation for D, the measure of expansion in the Universe; the lower four lines present numerical values for the expansion, density, radius, and age of the Universe. Einstein was offered a professorship at Oxford, but he found its refined formality oppressive, especially compared to the freedom and excitement of America where he eventually settled. Einstein, the consummate theoretician, was equally at home in a Universe of blackboards or backs of envelopes: while touring Mount Wilson Observatory with Edwin Hubble, Einstein's wife Elsa was told how the telescope was used to explore the structure of the Universe, to which she replied, "Well, well ... my husband does that on the back of an old envelope." (*Photograph by the author*)

roughly three-quarters of the "stuff" of the Universe.[15] Author Amir Aczel contends that Einstein's equation with the cosmological constant is our best approximation of what he calls "God's equation" [14]: the ultimate summary of how the Universe works. Whereas mass is the source of attractive gravity, for Einstein empty space is the source of repulsive energy. Einstein's blunders were often more fascinating and complex than even the triumphs of lesser scientists.

Later in 1917 the Dutch astronomer Willem de Sitter (1872–1934) found a solution to Einstein's general relativistic field equations that describes an (embarrassingly) empty and, as a result of the cosmic repulsion term, expanding Universe, the first prediction of a general cosmic expansion. Although the absence of matter was, of course, problematic, de Sitter's solution, for which test particles sprinkled into his model Universe recede from each other with a velocity proportional to their separation, soon became popular, particularly in view of its connection with the galactic redshifts that were being reported at the time. (Hubble referred to the redshift of the galaxies as the "de Sitter effect.") Throughout the 1920s cosmologists investigated which of the two relativistic alternatives, neither of which seemed to represent the real Universe, was the most satisfactory.

In 1922 the Russian mathematician and meteorologist Aleksandr Friedmann (1888–1925 from the aftereffects of dangerous high-altitude meteorological balloon flights) published a thorough analysis of the solutions to Einstein's cosmological field equations, showing that the static Einstein and the dynamic de Sitter solutions were merely two special cases of a more general solution which allowed for a dynamic Universe either expanding or contracting depending on its total mass-energy density. Five years later the Belgian astronomer-priest Abbé Georges Lemaître (1894–1966; Fig. 2.15), a

---

[15] While examining the Coma cluster of galaxies in 1933, the cantankerous Bulgarian-born Swiss astronomer Fritz Zwicky (1898–1974), who worked most of his life in the USA, was the first to use the *virial theorem*—essentially a statement of the balance between the tendency for member galaxies to fly away (that is, the kinetic energy of expansion), and the tendency for the mutual gravitational attraction of member galaxies to clump all together (i.e., the gravitational energy of collapse)—to infer the existence of unseen dark matter, the nature of which, like the more recently discovered mysterious dark energy, is still unknown. He found that the Coma cluster is nearly a hundred times more massive than the mass expected from the luminosity of its member galaxies. In 1937 he suggested that galaxy clusters could act as gravitational lenses, with their mass—seen or unseen—bending background light like a lens in a manner predicted by Einstein, an effect first observed in 1979 and soon followed by many more. The best models today suggest that ordinary (baryonic) matter accounts for only about five percent of the mass-energy of the Universe, just the top of the tip of the cosmic iceberg, with dark matter and energy amounting to about twenty-five and seventy percent, respectively. Stars, it would seem, are just for decoration, like a thin layer of white icing on a large, dark, chocolate, cosmic cake, showing, once again, that things are not always as they might first appear.

student of the pioneer English astrophysicist Sir Arthur Eddington (1882–1944) at Cambridge before studying at Harvard and MIT in America, independently arrived at the same equations which became known as the "Friedmann equations." But whereas Friedmann had approached the problem principally as a mathematical curiosity, relatively unconcerned about any relevance to the real Universe, Lemaître argued on physical grounds that the Universe *really* is expanding, and, using Slipher's redshifts and estimated distances, demonstrated that the recessional velocities of galaxies are proportional to their distances, just as Hubble announced two years later. (Nearly a century later, in 2018, the International Astronomical Union renamed Hubble's Law as the *Hubble-Lemaître Law* in recognition of Lemaître's contribution to the discovery of cosmic expansion.) In 1928 the American mathematician and physicist Howard Robertson (1903–1961) independently discovered the velocity-distance prediction of de Sitter's model and pointed to Hubble's accumulating data as its empirical confirmation. Nevertheless, the theoretical speculations of de Sitter, Friedmann, Lemaître, and Robertson were ignored until after Hubble's announcement of the velocity-distance relationship. Besides, an expanding Universe implied a beginning, a concept that many—scientists, in particular—found repulsive. "As a scientist I simply do not believe that the present order of things started off with a bang," Eddington declared in his 1927 Gifford Lectures. He shared the view of most others when, four years later, he added that "philosophically, the notion of a beginning of the present order of Nature is repugnant to me."

The Universe according to Lemaître's 1927 model was expanding but it did not have a beginning in time. In a remarkable precursor of the modern Big Bang theory of the origin of the Universe accepted today, Lemaître published a brief 457-word note in 1931 significantly titled "The Beginning of the World from the Point of View of Quantum Theory," in which he introduced quantum physics into cosmology, albeit in a rather vague and speculative style more in the manner of a visionary piece of "cosmopoetry" than a conventional scientific communication. He suggested an absolute zero of space and time, that the Universe of space and time originated about 10 billion years ago in a kind of explosive radioactive decay of a "primeval atom," a single "superquantum" containing the entire mass of the Universe in a region about 30 times larger than the Sun (and thus different from the modern Big Bang theory with its initial singularity of infinite density creating matter and energy not within preexisting space and time, but creating both space and time as well) that expanded and decayed radioactively into more and more quanta of less and less energy, with the interconversion of light and matter occurring in accordance with the laws of relativity and quantum physics. Cosmic

Fig. 2.15 Issued on the 100th anniversary of his birth, these Belgian postage stamps honor the Belgian astronomer-priest Abbé Georges Lemaître, who in 1927 predicted Hubble's (1929) Law governing the expansion of the Universe, and in 1931 predicted the origin of the Universe from an exploding "primeval atom," a remarkable precursor of today's Big Bang theory. Lemaître is rightly honored as the father of Big Bang cosmology and hence of modern cosmology itself. (*Photograph by the author from author's collection; courtesy of Louis and Mona van Holsbeeck*)

rays, Lemaître (erroneously) suggested, are the remnants of the original explosion, "ashes and smoke of bright but very rapid fireworks." This was the first of what would later be labeled (pejoratively, as were the terms "Impressionism" and "Gothic" in the art world of previous centuries) the *Big Bang* model of the Universe (Fig. 2.16), "generally considered," quoting science historians Helge Kragh and Dominique Lambert, "to be one of the most important events in the history of cosmology ever, easily comparable with… the Copernican revolution" [15, p. 445].

Though it was initially largely ignored, by the late 1930s the idea of an evolutionary Universe having a definite, finite age, and described by the laws of general relativity, gained increasing respectability. While the notion of a beginning and, consequently, of a Creation was conceptually problematic—creation by whom or by what? (Einstein warned Lemaître that his hypothesis "smells too much of creation")—and was therefore claimed to be more metaphysical or theological than scientific, lacking as it did any observational support, it must have resonated, along with Lemaître's persistent Grosseteste-like fascination with light, with his Catholic belief in the Bible and the Genesis *Fiat lux* story of Creation. (The medieval English philosopher-bishop Robert Grosseteste described the primordial, divinely created Universe as a point of light which instantaneously propagated itself into an expanding sphere and eventually gave rise to the celestial bodies.) Nevertheless, although

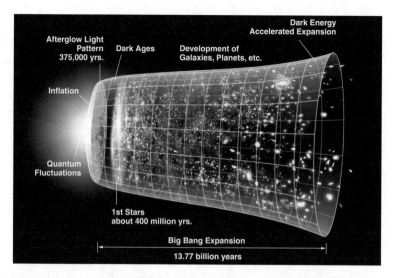

Fig. 2.16 An illustration of the evolutionary history of the Universe. Time advances along the horizontal axis, while space is measured vertically. Note the period of rapid inflation initiated by the Big Bang. The "Dark Ages" refers to the time before the formation of the first stars. See Fig. 2.22 for more on the "Afterglow Light Pattern." As a long-time local science fair judge, one of my favorite science cartoons, this one by Nick Downes, shows a boy with detonator under arm, pulling his little wagon loaded with dynamite and a sign reading "THE BIG BANG" into a grade-school science fair populated with the usual mundane—and decidedly safer—projects on photosynthesis, crystal growth, baking soda and vinegar volcanoes, and model solar systems. (*Wikimedia Commons, public domain*)

he insisted that the primeval atom was created by God *ex nihilo,* Lemaître otherwise avoided a theological interpretation of his work in cosmology.[16]

There was another problem with theories of the Universe like Lemaître's that postulated an origin in time: such a Universe is embarrassingly younger than the known age of Earth. If, as Hubble's observations suggested, the Universe is expanding, the age of the Universe—that is, the time since the expansion began—could be calculated by simply running the cosmic clock backwards until the Universe was back in its original highly compressed state. Since the Hubble constant H_0 gives the expansion velocity per unit distance and is therefore a measure of the expansion rate, the age of the Universe—the

---

[16] In a 1951 address to the Pontifical Academy of Sciences, headed at that time by Lemaître, Pope Pius XII endorsed Big Bang cosmology, welcoming the corroboration of the Church's doctrine of Creation—*creatio ex nihilo*—by the tenets of modern science. As the pope pontificated, "the mighty *Fiat* of the Creating Spirit billions of years ago [created] with a gesture of generous love matter bursting with energy." In the opinion of British physicist and theologian John Polkinghorne, "Such a Papal endorsement of a particular scientific theory was as embarrassingly misconceived as was the opposition by his predecessor Urban VIII to the ideas of Copernicus and Galileo." Indeed, Lemaître worked hard to quash any further inappropriate mixing of science and religion.

so-called *"Hubble time"*—is, from Hubble's Law, the reciprocal of H_o: *time = distance/velocity* $= 1/H_o$. Hubble's original value of $H_o \approx 500$ km/s per Mpc (recall 1 Mpc = 1 megaparsec = 3.26 million light years) gives a rather uncomfortably young age for the Universe of only 2 billion years, younger than the known age of Earth at that time. (The presence on Earth of radioactive *isotopes*—each of two or more forms of the same element containing different numbers of neutrons in their nuclei—with half-lives comparable to this Hubble time was a motivating factor in Lemaître's proposed model for the origin of the Universe.) It was certainly difficult to understand how Earth could be older than the Universe, and this problem was much discussed throughout the 1930s and 1940s.

One solution to the timescale problem was the so-called *steady state cosmology*, a model of the Universe proposed in 1948 by a trio of Cambridge University graduates, Fred Hoyle (1915–2001; see Fig. 3.15), Hermann Bondi (1919–2005), and Thomas Gold (1920–2004). Twentieth-century cosmological equivalents of earlier uniformitarian geologists, steady state cosmologists answered cosmic catastrophists like Lemaître just as James Hutton had answered late-eighteenth-century geological catastrophists in his time. Like Hutton they saw "no Vestige of a Beginning—no prospect of an End" on the cosmic horizon, arguing that matter was continuously created to fill in the voids left by receding galaxies in an infinitely old Universe that looked the same for all time (hence steady state), past, present, and future. A steady state Universe requiring neither a definite beginning nor a foreseeable end was Hutton's uniformitarianism writ BIG. It was Hoyle, an ardent atheist, who in a 1949 BBC radio program coined the term "Big Bang" in reference to "the hypothesis that all the matter in the universe was created in one big bang at a particular time in the remote past." The proposition of the continuous creation of matter aroused much debate (for example, where does it come from?) and led many to brand the theory "unscientific romanticizing" or "science-fiction cosmology."

One strong argument *for* the steady state theory, if only because it was a strong argument *against* the Big Bang, was Hoyle's work in the 1950s on stellar nucleosynthesis to be discussed in Chap. 3, which was motivated largely by the inability of cosmology, particularly early Big Bang models, to address the problem of the cosmic abundance of the elements. Lemaître's realization that the birth of the Universe was accompanied by very high temperatures happened to coincide with the birth of nuclear physics; he remarked that the study of the origin of the Universe "is atomic physics on a large scale." The fusion of the two, which incidentally brought a new and needed dose of scientific respectability to cosmology, resulted in attempts to model chemical element synthesis in the early Universe via various high-energy nuclear processes, a field of investigation that became known as *Big Bang nucleosynthesis* to distinguish it from stellar nucleosynthesis. By 1946 the Russian émigré George Gamow (1904–1965),[17] who brought quantum physics to nuclear physics two decades earlier with his theory of α-decay ("*alpha-decay*," a nuclear decay process whereby a nucleus emits an *α-particle*, two protons and two neutrons bound together into a composite particle identical to the nucleus of a helium atom), presented a revised Big Bang model that combined nuclear physics in the very early Universe with the Friedmann equations, indicating how the chemical elements could form during the earliest phase of cosmic expansion when the Universe was still very hot.

Gamow's model was substantially modified and refined two years later in collaboration with his student Ralph Alpher (1921–2007) and Alpher's colleague Robert Herman (1922–1997) using newly unclassified neutron-capture cross sections (the likelihood of interaction between an incident neutron and a target nucleus), sensitive information for World War II's atomic bomb project. However, it was soon realized that their theory, attempting to build up all the elements from successive neutron capture, was unable to explain the production and abundance of elements heavier than helium, but rather surprisingly got just about the right abundance for helium and predicted a present-day background temperature of 5 K as the early, hot Universe cooled, although it never occurred to them that it might be observable. ("K" denotes the absolute Kelvin temperature which equals the Celsius temperature plus approximately 273; e.g., water freezes at 0 °C = 273 K and boils at 100 °C = 373 K, with "absolute" zero being −273 °C.) Along with the development of stellar nucleosynthesis by Hoyle's group, it became clear that the physical conditions deep within stars, unlike those in an expanding—and hence cooling Universe of decreasing density—could bridge the gaps at atomic mass numbers 5 and 8, for which no stable nuclei exist, and thus successfully account for the abundances of nearly all the elements. This was widely hailed as a strong argument against the Big Bang theory despite the fact that the steady state theory could not even address element production except, like the creation of matter itself, in a totally ad hoc manner.

---

[17] Gamow, a well-known prankster and author of several popular physics books for the general reader, was purportedly ticketed for running a redlight in downtown Boulder when he held a professorship at the University of Colorado. He went to court and told the judge that because he was approaching the traffic light, the red color was Doppler shifted to green. The judge, who was evidently—if uncharacteristically—familiar with the physics, quickly estimated that in order for Gamow to see a red light as green, the professor would have been traveling towards it at a speed of nearly 30% the speed of light. He gave Gamow the choice of paying either the $25 fine for running the redlight or a (much, MUCH higher) fine of $1 for every mile per hour he was moving over the posted speed limit of 25 mph. Gamow paid the $25.

The 1960s marked a turning point for observational cosmology and the Big Bang theory (even if the TV series named after the latter had to wait until the early years of the next century to debut), and already in the 1950s the value of Hubble's constant was revised downward giving a much more comfortable age for the Universe of roughly 10 billion years; today's best estimate is 13.8 billion years (recall Fig. 2.16). Progressive increases in the estimated age of the Universe, like the earlier appreciation of the increasing age of Earth, progressively undermined our ancient anthropocentric perceptions. By 1960, radio astronomers realized that either the number or the luminosity of faint and thus presumably distant and therefore (allowing for the light travel time, remembering that we see distant objects as they were at the time the light left them) young radio sources were less than expected for a Universe in a steady, nonevolving state. Five years later it became clear that counts of newly discovered *quasars*—bright, point-like sources now known to be the highly energetic cores of young galaxies harboring supermassive black holes[18]—also contradicted steady state theory: *all* quasars exhibit large redshifts and hence, interpreted via Hubble's equation, are *all* far away in space and therefore far back in time. Contrary to steady state theory, the Universe therefore looks different at different times in its history: in a steady state Universe, quasars should be distributed uniformly, some close, some far.

The final nail in the steady state coffin was the serendipitous detection in 1964 of the cosmic microwave background (CMB), the cooled remnant of the first light that could travel freely throughout the now transparent Universe after the temperature dropped below 3000 K allowing previously free electrons and protons to combine forming neutral atoms no longer capable of scattering radiation (whence the "surface of last scattering" marking the decoupling of radiation from matter—the "Afterglow Light Pattern" shown in Fig. 2.16 occurring some 380,000 years after the Big Bang when the "fog" cleared, the most ancient signal ever recorded). While modeling Big Bang nucleosynthesis in the late 1940s, Gamow and his group realized that the early Universe, hot as it was, must have been dominated by radiation rather than matter, and that as a result of universal expansion this relic radiation should have cooled down to about 5–10 K.[19] The spectrum of thermal

---

[18] The word is an acronym for *quasi-stellar radio sources*, although today we know that less than 10% of these objects are strong radio emitters: hence their new label, QSOs for *quasi-stellar objects*.

[19] Already in 1941, the astronomer Andrew McKellar announced that the temperature of space is about 3 K based on his study of narrow absorption features in the spectra of stars due to interstellar cyanogen (C_2N_2, the first molecule detected in the interstellar medium), but no one connected McKellar's measurement with the possibility of detecting remnant radiation from the Big Bang. In the days before the switch to digital television broadcasts, about one percent of the static on televisions found between channels was caused by this relic radiation. I certainly saw my share as a kid.

radiation at this low temperature peaks in the microwave (short-wave radio) portion of the *electromagnetic* (EM) *spectrum* (Fig. 2.17). But their prediction was ignored and eventually forgotten until Arno Penzias (1933–2024), a child refuge from Nazi Germany, and Robert Wilson (b. 1936), working together at Bell Labs in New Jersey with a horn antenna that was used to conduct experiments testing microwave relay communication via artificial satellites (Fig. 2.18), detected, quite by accident, an unexplained, isotropic "excess antenna temperature" (even after excluding the effect of "a white dielectric material"—pigeon poop—coating the antenna throat) characteristic of a 3.5 K *blackbody*, a material that emits a spectrum of electromagnetic radiation dependent solely on its temperature (hence "thermal" radiation).

The temperature of blackbody radiation is inversely proportional to its "typical" wavelength—the wavelength ("color") corresponding to the peak intensity of the radiation—a Nobel Prize-winning discovery made in 1893

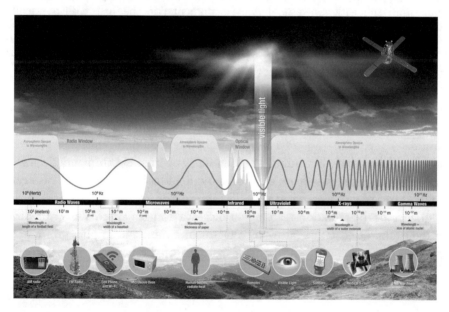

Fig. 2.17 The electromagnetic (EM) spectrum, stretching from the long-wavelength (low-frequency) radio waves to the short-wavelength (and hence high-frequency) gamma rays, with relevant sources/uses positioned beneath each portion along the bottom of the figure, and the atmospheric transparency indicated across the middle of the figure. Note the relatively narrow spectral range of visible light, which runs from about 400 nm for blue light to about 700 nm for red light (1 nm—"nanometer"—equals one billionth of a meter, 10^{-9} m). Because all EM waves travel in a vacuum at the same speed c, the speed of light, wavelength λ is inversely proportional to wave frequence f (measured in Hz, i.e., cycles per second): $c = \lambda\, f$. (*Photograph courtesy of NASA, public domain*)

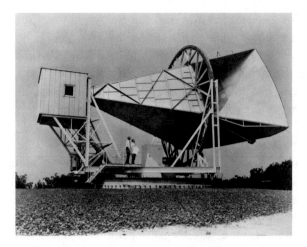

Fig. 2.18 The horn antenna used by Arno Penzias and Robert Wilson at Bell Labs in their serendipitous discovery of the cosmic microwave background, remnant radiation from the early Universe. It was instrumental in the first passive satellite communication experiment when, in 1960, it received microwaves bounced off the 100-foot-diameter metallized Mylar balloon satellite, *Echo*, that were transmitted from Goldstone, California, a concept first proposed by science fiction author Arthur C. Clarke in 1945. (*Wikimedia Commons, public domain*)

by the German physicist Wilhem Wien (1864–1928) and known as *Wien's displacement law* ($\lambda_{\text{peak}} \sim 1/T$, denoting wavelength with the Greek letter lambda, λ; see Fig. 3.9). *Photons* of blackbody radiation—massless elementary particles that are bundles ("quanta") of the electromagnetic field, including, as is the case here, electromagnetic radiation such as light (Greek, phōs; Latin, photos), carrying a specific amount of energy proportional to the frequence of the radiation—are one typical wavelength apart. Thus, the drop in the background temperature by a factor of a thousand implies an increase in photon wavelength—and hence expansion of the Universe—of an equal factor of a thousand.[20]

---

[20] "Red hot" is a misnomer: objects emitting more thermal radiation in the short-wavelength blue region of the spectrum are hotter. The *bluish* white-hot star Sirius, the brightest star in the sky after the Sun, for example, has a surface temperature (10,000 K) nearly twice that of our *yellowish* Sun (6000 K), which in turn is nearly twice as hot as the *red* giant star Betelgeuse (3600 K) in the constellation of Orion. One wonders why meteorologists use blue to depict cooler temperatures and red for warmer regions. Outside the visible spectrum, the detection of thermal X-rays which, as Fig. 2.17 shows, have very short wavelengths on the order of the size of an atom, indicates very high temperatures in excess of a million degrees, such as are associated with black-hole accretion disks, while at the other end of the spectrum, the background radiation of a Universe just a few degrees above absolute zero peaks, as we have just noted, in the longer wavelength microwave radio portion of the spectrum. Wien's displacement law is λ_{peak} (cm) $\simeq 0.3/T$, where the temperature is measured in kelvins.

Although Penzias and Wilson did not connect the anomaly they discovered with cosmological theory (they brought it to the notice of the astronomical community with a brief two-page "Letter to the Editor" in *The Astrophysical Journal*), a group of Princeton physicists just down the road, including Robert Dicke (1916–1997) and James Peebles (b. 1933), had just reworked the theory and constructed their own antenna for the purpose of detecting this very radiation. They immediately recognized that they were scooped by Penzias and Wilson who had unknowingly made an immensely important cosmological discovery: the cosmic background radiation left over from the Big Bang, a discovery for which they were awarded half the 1978 physics Nobel Prize. In 1966 Peebles, who was awarded a share of the 2019 physics Nobel Prize "for theoretical discoveries in physical cosmology," calculated a cosmic helium abundance of 27% based on Big Bang assumptions and a cosmic background temperature of 3 K, in excellent agreement with observations and thus yet another triumph of the new Big Bang cosmology. No longer could the existence of a hot Big Bang origin for the Universe be denied.

With the elimination of its steady state rival—recounted here as an example of how science works and progresses, testing hypotheses with observations—Big Bang evolutionary theory rose to paradigmatic status and cosmology itself gained scientific respectability throughout the general scientific community. Until the empirical results of the 1960s, cosmological theory was as often philosophical as scientific, and, as the Nobel Prize-winning physicist Steven Weinberg admitted [16, p. 4], "the study of the early universe was widely regarded as not the sort of thing to which a respectable scientist would devote his time." As Toulmin and Goodfield pointed out in 1965 [17, p. 262], it was the case that "[i]n all other historical sciences [except cosmology], the crucial transition to the phase of cumulative advance has been marked by the changeover from a priori patterns of theory to an empirical, developmental method of enquiry."[21]

Discoveries in the rapidly developing field of high-energy particle physics in the 1960s and 1970s, together with increasing confidence in the validity of hot Big Bang cosmology following the discovery of the CMB, led to the realization of a close connection between particle physics and cosmology, with the early Universe now appreciated as nature's particle accelerator reaching energies far in excess of those accessible in terrestrial facilities where,

---

[21] Nevertheless, some still question the field's scientific status, linking it, as it had been for most of history, more closely with speculative philosophy than with modern science. In 2007 the astronomer Michael Disney published an article in *American Scientist* titled "Modern Cosmology: Science or Folktale?" emphasizing what he perceives to be rather flimsy observational support for the current Big Bang scenario. Cosmology is not only a highly mathematical science; it also relates, by the very scope of its reach, to deep philosophical and theological questions.

in the interest of furthering civilization, physicists smash subatomic particles together (whence the term "atom smasher") and look for interesting physics in the debris. In this new symbiotic relationship known as *particle cosmology*—a modern-day equivalent of the ancient microcosm-macrocosm mix (recall Figs. 1.2 and 2.1, and see Fig. 5.1 and Note 1 in Chap. 5)—cosmological ideas can be tested in the laboratory and, conversely, theories of elementary particle physics can be tested on the largest imaginable scale of the Universe itself. For example, detailed calculations based solely on the Big Bang theory limiting the different types of neutrinos to no more than three were later confirmed by particle accelerator experiments at CERN and served to further increase confidence in the correctness of the Big Bang model.[22]

In 1980 the American physicist Alan Guth (b. 1947; Fig. 2.19) announced his variation, the first of any consequence, of Big Bang theory called *inflationary cosmology*, which brought particle physics to bear on several outstanding problems within the standard Big Bang model, in particular the so-called "fine-tuning" problems—why the Universe is so Rudyard Kipling-like "just so." Why, for example, is the Universe so "smooth" (i.e., homogeneous and isotropic, the same everywhere and in every direction)—with the CMB exhibiting very little background temperature fluctuation (only one part in 100,000; $\Delta T/T \sim 10^{-5}$, using power-of-ten scientific notation with the exponent indicating the number of times 10 is multiplied by itself, here 5 times), but just enough associated density inhomogeneity to seed the formation of galaxies and galaxy clusters in an ever-expanding Universe? And why is it so "flat" (the "*flatness problem*")—having a critical density such that its kinetic energy of expansion is perfectly balanced by its gravitational energy

---

[22] In the early twentieth century, the atom became the microcosmic equivalent of the Solar System, in this case, tiny electrons encircling a much more massive central nucleus rather than tiny planets orbiting the Sun, a model that facilitated our understanding of atomic physics. In yet another, but related, example of the meshing of microcosm and macrocosm, astronomer John Gribbin comments on "a very profound discovery" [18, p. 599]:

> The age of the Universe is essentially calculated from the general theory of relativity, and deals with the laws of physics on the very large scale; the ages of stars are . . . essentially calculated from the laws of quantum mechanics, physics on the very small scale. Yet the age of the Universe comes out to be just enough older than the ages of the oldest stars to allow the time required for the first stars to form after the Big Bang [recall Fig. 2.16]. This agreement between physics on the largest and smallest scales is an important indication that the whole of science is built on solid foundations.

"Astronomy affords the most extensive example of the connection of the physical sciences" the Scottish polymath Mary Somerville wrote in her 1834 *On the Connexion of the Physical Sciences*, one of the best-selling science books of the nineteenth century and one of the first popular science books. In a review of the book in March 1834, the Cambridge polymath and wordsmith, William Whewell, coined the word "scientist" on the same basis as "artist."

of contraction, a balance today that implies an incredibly finely tuned early Universe, balanced to about one part in 10^{60} before inflation? (This is a HUGE number, 1 followed by 60 zeros!—thus the convenience of scientific notation for the very large and very small numbers often encountered in science.)

In Guth's model, the very young Universe underwent extreme supercooling due to a symmetry-breaking phase transition in a *false cosmic vacuum*—a relative energy minimum above the lowest possible energy (are you getting all of this?)—with an accompanying release of energy, much like the condensation of steam to liquid water or the freezing of water to ice releases pent-up latent heat, expanding the Universe suddenly by a huge factor of about 10^{30} within a time of less than 10^{-30} s (recall Fig. 2.16). This explains the flatness of the Universe, much as the surface of a huge balloon undergoing such a tremendous expansion would appear flat (or why Earth, large as it is, appears flat), as well as its smoothness: distant regions in the Universe not causally connected today (i.e., outside each other's observable horizon, whence the "*horizon problem*") and hence unable to exchange information such as temperature and density because the light travel time between them, a measure of the fastest means of exchanging information, exceeds the age of the Universe, would have been much closer and hence causally connected before inflation. (Inflation also solves a third problem, accounting for the apparent absence in the Universe today of *magnetic monopoles*, hypothetical particles carrying a single magnetic pole predicted by quantum theory nearly

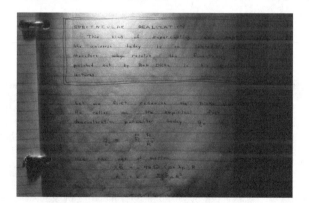

Fig. 2.19 Detail of Alan Guth's notebook on display at Chicago's Adler Planetarium, leading to his proposal of inflationary cosmology. Note the comment boxed at the top with the heading "SPECTACULAR REALIZATION" that a rapid, early inflation would, as he writes three lines down, "resolve the fine-tuning problems" of the flatness and smoothness of the Universe today. (*Photograph by the author*)

a century ago: the rapid thinning out of the early Universe during inflation makes them essentially undetectable today.)

During *symmetry breaking*, previously unified fundamental forces "break" apart (separate) into the distinct forces recognized today. Within the framework of inflationary cosmology, the early Universe thus becomes a testing ground for models of particle physics including the anatomically sounding GUT (*Grand Unified Theory*: a model unifying the electromagnetic, weak, and strong forces of the so-called *Standard Model of particle physics* into a single force at high energies) and ToE (*Theory of Everything*: GUTs with gravity, such as is modeled by *string theory* in which particles are imagined as tiny vibrating strands of energy).[23]

For reasons not yet understood (exotic particles that violated conservation laws?), the breaking of symmetries involving matter–antimatter interchange during the early Universe resulted in a slight excess of matter over antimatter amounting to about one billion and one matter particles for every billion *antimatter* particles, oppositely charged counterparts of ordinary matter— Guth calls this the "ultimate free lunch"—leaving the Universe today with about a billion photons, the product of matter–antimatter annihilation, for every surviving nucleon. "So," as authors Philip Dauber and Richard Muller point out [19, p. 166]—and as we shall see in the next chapter—"not only are we made of matter cooked in stars, the ingredients from which the stars formed were just a tiny dash of debris left over from a universe once a billion times more massive." Indeed, if there were equal amounts—a perfect balance—of matter and antimatter, the Universe would be all radiation, as all the matter and antimatter would have long ago annihilated—and *we wouldn't be here!*

---

23 The *electromagnetic* force is responsible for all things electrical and magnetic, including light, and, like *gravity*, is a long-range force, the effects of which can be noticed directly in everyday life, but, unlike gravity, which is always attractive, can be either attractive or repulsive. The *weak* and the *strong* forces are short-range forces acting only over subatomic distances of order 10^{-15} m—about the size of a nucleon, 100 million times smaller than the width of a human hair—that govern nuclear interactions inside atoms; the *nuclear force* is a residual effect of the strong interaction that binds atomic nuclei together, and the weak force is responsible for the radioactive β-decay of atomic nuclei. The gravitational force, first described mathematically by Isaac Newton in the late seventeenth century, is more generally attributed via Einstein's general theory of relativity to the curvature of spacetime; the other three forces can be described by quantum fields mediated by elementary particles (such as the photon for the electromagnetic interaction) described by the Standard Model of particle physics; there is not yet a quantum theory of gravity, the "Holy Grail" of physics. As inferred by its name, the strong force is the strongest of the four fundamental forces, about 100 times stronger than the electromagnetic force (which is why it is able to hold together atomic nuclei against the electrical repulsion of constituent protons), a million times stronger than the weak force, and 100 trillion trillion trillion (10^{38}) times stronger than gravity. Although gravity is the weakest of the forces, it is the dominant force over astronomical scales where one encounters very large masses (as in planets, stars, and galaxies) with no net electric charge.

This number, the 10^9 photons-to-nucleons number, crucial to cosmological models, can be worked out from observations of the CMB. Using the Planck radiation law derived in 1900 by the German physicist and father of quantum theory, Max Planck (1858–1947), one finds the number density of photons in the Universe today to be about 500,000 per liter (or about a thousand photons per teaspoon, one teaspoon holding a volume of about 5 cm^3) for its current background temperature of 2.7 K [see, e.g., 16, p. 171–173]. The so-called critical density of the Universe—the density required for the observed "flatness"—is about 9×10^{-30} g/cm^3,[24] so the density of baryonic matter (75% hydrogen at 1 g/mole and 25% helium at 4 g/mole), which accounts for just under 5% of that, is (recalling Avogadro's number of 6.02×10^{23} particles per mole) about one baryon per *ten thousand* liters (or about one per *one million* teaspoons, which can be compared to the air you're breathing, which amounts to about *one hundred million million million* molecules of oxygen and nitrogen per level teaspoon!), giving a photon-to-baryon ratio of about 10^9, a billion photons for every nucleon. Indeed, there must have been a very large ratio of photons to nucleons to prevent the "cooking" of all the hydrogen into helium in the early Universe.

2.3   Big Bang Nucleosynthesis

There is surely a peece of Divinity in us, something that was before the Elements, and owes no homage unto the Sun.
— British natural philosopher and physician Thomas Brown,
Religio Medici (1642), Pt 2, Sec 11

In a curious sense, cosmological nucleosynthesis is a relatively easy problem…. The staggering fact is that this simple theory works so well, on such a grand scale.
— astronomer David Arnett in *Supernovae and Nucleosynthesis: An Investigation of the History of Matter from the Big Bang to the Present* [20, p. 119]

---

[24] Weinberg [16, pp. 167–168] shows how the critical density can be found in a very straightforward manner by balancing the kinetic energy of expansion with the gravitational energy of contraction. If the average density of the Universe were greater than this critical value, the total mass-energy of the Universe would eventually halt the expansion and lead to a subsequent contraction, and ultimately to a "Big Crunch" when everything comes together into a singularity of infinite density, perhaps to explode again in another Big Bang. In this case, the Universe is said to be *closed*, and can be likened to throwing a ball up with less than its escape speed: it will reach a maximum height, determined by its initial speed (higher for greater speeds), and then fall back to Earth. On the other hand, if the average density were less than the critical density, the Universe will continue to expand forever, possibly resulting in a "Big Rip." This is the situation for an *open* Universe, and would be like a ball thrown upward with a speed greater than its escape speed, never to return again.

Having traced the historical development of our understanding of the origin and evolution of the Universe, hot enough in its early history, and thus certainly capable of supporting various high-energy nuclear processes, we now investigate the details of Big Bang nucleosynthesis, the synthesis of chemical elements in the early Universe. The observed 3 K temperature of the CMB today, together with the inferred mass-energy density of the Universe allow us to predict the observed cosmic abundances of the lightest elements. Indeed, the abundances of hydrogen and helium *require* a Big Bang origin for the Universe and a present-day CMB. As we shall see, tracing important eras and events in the history of the Universe, the physics of the very large—the Universe and, by definition, everything in it—was determined long ago by the physics of the very small. All things really are connected.

$$\text{Planck era: } t = 10^{-43} \text{ s; } T = 10^{32} \text{ K}$$

We begin at the beginning, some 13.8 billion years ago. (The reader may wish to refer to Fig. 2.20 for a collective visualization of the important timesteps and associated processes summarized in the following few pages.) Amazingly, we can trace the history of the Universe back to the *Planck time*, a mere 10^{-43} s after the singularity we call the Big Bang, the very beginning of our Universe.[25] At the Planck scale, where the quantum effects of gravity are expected to dominate, quantum field theory and general relativity, along with the predictions of the Standard Model of particle physics—the theory describing all known elementary particles and three of the four known fundamental forces (the electromagnetic and the strong and weak nuclear forces, but excluding a complete theory of gravity as described by general relativity; recall Note 23)—are not expected to apply. Because we do not yet have a theory of quantum gravity—a "Theory of Everything" (ToE)—we cannot say anything scientifically about what occurred before the Planck time (but you can ask your favorite philosopher, theologian—or Republican—for their opinion on such matters).

$$\text{Inflation: } t = 10^{-32} \text{ s; } T = 10^{27} \text{ K}$$

---

[25] Originally proposed in 1899 by Max Planck, *Planck units* are a natural system of units of measurement defined using fundamental properties of nature rather than those of a particular prototype object (such as the size of a king's foot), exclusively in terms of four universal physical constants: c, the speed of light; G, the Newtonian gravitational constant; $\hbar$, the reduced Planck constant (i.e., the Planck constant h divided by 2π); and k, the Boltzmann constant. The *Planck temperature* 10^{32} K is the temperature of the Universe at the Planck time; the wavelength of a thermal photon at this temperature is equal to the *Planck length* 10^{-35} m. The *Planck mass*, about 2×10^{-8} kg, equivalent (via Einstein's famous mass-energy equivalence formula, $E = mc^2$) to a *Planck energy* of about 10^{19} GeV, is the fourth and final Planck unit.

From the Planck time until the end of inflation some 10^{-32} s after the Big Bang, the temperature was so high (about 10^{27} K at the end of inflation, Zel'dovich's "glowing hell") that GUTs are needed to understand the physics taking place because the electromagnetic and the strong and weak forces are indistinguishable ("unified") at these high energies for which differences disappear between quarks (recall Note 10) and leptons (recall Note 5 of Chap. 1).

As the temperature of the expanding Universe dropped (think of the cooling experienced by a rapidly expanding gas), the temperature and hence energy dropped below the mass-energy (mc^2) of the exchange particles that mediate the various fundamental forces, which then separate out into distinct forces breaking the previous symmetrical (unified) state. At the end of inflation, the Universe had "cooled" to the point where the average energy is no longer much greater than that of the carriers of the combined electromagnetic and strong and weak nuclear forces, an energy estimated to be about 10^{14} GeV (= kT, where $k = 1.38 \times 10^{-23}$ J/K is the Boltzmann constant and the temperature $T = 10^{27}$ K; for comparison, the mass-energy of the proton and the neutron is just under 1 GeV and that of the electron is about 0.5 MeV; note that for the very high temperatures realized during the early Universe, the temperature measured in kelvins is essentially the same in degrees Celsius since the two scales differ by only 273°). Accordingly, the strong force separated out and became distinct from the *electroweak* interaction, a unified description of the electromagnetic and weak forces discovered in the second half of the twentieth century (see the next timestep for more details). Here, the prefixes "G" and "M" stand for "Giga" and "Mega" and

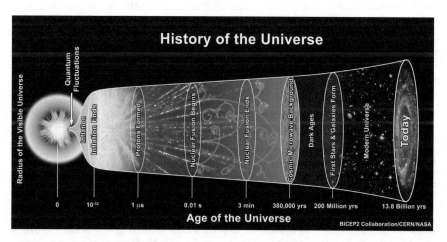

Fig. 2.20 Important milestones discussed in the text occurring during the 13.8-billion-year "History of the Universe." (*Wikimedia Commons, public domain*)

equal one billion (10^9) and one million (10^6), respectively, and one *electron volt*, abbreviated eV, is the energy an electron has when accelerated by a potential difference of one volt and is equal to 1.6×10^{-19} J (joules, the metric unit of energy, not a lot of energy inasmuch as a 100-W lightbulb emits 100 J every second; a power of 1 W = 1 J/s).

$$\text{Electroweak transition: } t = 10^{-12} \text{ s}; T = 10^{15} \text{ K}$$

By 10^{-12} s, the Universe cooled to a point where the unified electroweak force separated into distinct electromagnetic and weak forces because the average energy is comparable to the mass-energy of the W and Z particles (about 80 GeV and 90 GeV, respectively) that mediate the weak interaction. The Boltzmann energy-temperature equation ($E = kT$) gives an equivalent temperature of about 10^{15} K, hot enough still. Much like heads and tails cannot be distinguished while a tossed coin is in the air, making the coin in that sense symmetrical, before this symmetry breaking in the early Universe, the weak force was indistinguishable from and as strong as the electromagnetic force: the weak and electromagnetic interactions were two sides of the same coin in the high-energy early Universe, just as the electric and magnetic forces are two sides of the same electromagnetic coin today.[26] Afterwards, the weak force separated from and became distinct from and much weaker than the electromagnetic force, just as heads and tails are distinct at the symmetry-breaking end of a coin toss. Particle accelerators today, such as those at CERN that straddle the border of France and Switzerland near Geneva, are just barely capable of reaching such energies and have experimentally verified electroweak unification, as the kinetic energy of the colliding particles at these high energies is converted into the mass-energy of the W and Z particles via Einstein's famous mass-energy equivalence formula, $E = mc^2$. A collider the size of the orbit of Mars would be needed to reach the very high energies associated with GUTs.

By 10^{-6} s, the temperature and energy of the Universe dropped low enough ($T < 10^{12}$ K) that quarks did not have enough energy to remain free and were thereafter confined in *hadrons*, subatomic particles like the familiar nucleons (protons and neutrons) that participate in the strong nuclear force, a process

---

[26] The realization of the unity of electric and magnetic effects is due to the work of several nineteenth-century physicists, including Hans Christian Oersted, André-Marie Ampère, Michael Faraday, and James Clerk Maxwell. In the seventeenth century, Isaac Newton unified the celestial physics of Johannes Kepler with the terrestrial physics of Galileo, long set asunder by Aristotle, the first step towards GUTs and ToEs.

known as the quark-hadron phase transition.

Neutrino decoupling and electron-positron annihilation:
$$t = 1\,\text{s}; T = 10^{10}\,\text{K}$$

Before the Universe was one second "old," when temperatures exceeded 10^{10} K, radiation and matter were strongly coupled as a result of matter–antimatter creation ($\leftarrow$)-annihilation ($\rightarrow$):

$$\text{particle} + \text{antiparticle} \rightleftarrows \text{radiation} ,$$

where the radiation is manifested in high-energy gamma-ray photons because the annihilation of even the lightest particle-antiparticle pair—the electron-positron (e^+) pair ($e + e^+ \rightleftarrows \gamma + \gamma$)—results in photons (denoted with the Greek letter "gamma," γ) having energies at least as large as the rest mass-energy of the electron ($m_e c^2 = 0.511$ MeV) in order to conserve energy during the process, an energy in the gamma-ray portion of the EM spectrum (recall Fig. 2.17; by contrast, photons of visible light have energies of only a few eV).[27] Matter and radiation were therefore in thermal equilibrium with an equilibrium abundance of photons, electrons and positrons, neutrinos and antineutrinos, and protons and neutrons (10^9 times less abundant than the others), and the energy density (i.e., energy per unit volume) of radiation was much greater than that of matter in this still radiation-dominated Universe. Because neutrons are slightly more massive than protons ($[m_n - m_p]c^2 = 1.29$ MeV), more energy had to be borrowed from the hot plasma to make a neutron than a proton. The Boltzmann equation gives the equilibrium ratio of neutrons to protons:

$$N_n/N_p = \exp[-(m_n-m_p)c^2/kT] \text{ (Boltzmann equation)},$$

where "exp" denotes exponentiation: raising the Eulerian e ($= 2.718....$) to the power $-(m_n - m_p)c^2/kT$ (see, e.g., 21, p. 224—but *please* don't worry about the math, as we will use very little of it and only when absolutely necessary to calculate a quantitively important result: nevertheless, you should

---

[27] In one of his "miracle year" 1905 papers, the one that explained the photoelectric effect using a particle ("photon") rather than a wave description of light (and the one for which he was awarded the 1921 Nobel Prize in Physics), Einstein showed that the energy E of a photon is proportional to its frequency f, with the proportionality constant being Planck's constant h ($= 6.626 \times 10^{-34}$ J·s): $E = hf$. This gives a photon frequency of about 10^{20} Hz for the 0.511 MeV rest mass-energy of an electron (or positron since they have the same mass). As Fig. 2.17 shows, photons having this frequency are gamma rays.

appreciate the power of math as *the* language of the exact sciences). For T >> 10^{10} K, the numbers of protons and neutrons are essentially equal: $N_n/N_p \simeq 1$.

But at temperatures just above 10^{10} K, a thousand times the temperature at the center of the Sun (equivalent to a Boltzmann energy of about 1 MeV), the Universe became transparent to neutrinos, which were thereafter rarely absorbed by matter because their *mean free path*—the average distance they travel before interacting with matter—exceeded the size of the Universe at that time and forevermore (or, equivalently, the time scale for weak interactions became greater than the age of the Universe). From that point on, the weak interaction became unimportant ("*weak freezeout*") as neutrinos decoupled from matter ("*neutrino freezeout*"), just as, as discussed in the previous section, radiation decoupled from matter some 380,000 years after the Big Bang to produce the CMB when the Universe became transparent to radiation (see also that timestep below). There should thus be a cosmic neutrino background, just as there is a CMB, and because the threshold temperature for electron-positron creation-annihilation is about 6×10^9 K ($= m_e c^2/k$), a little lower than that for neutrino freezeout, there were still some electron–positron annihilations producing photons (and hence heat) after the neutrinos decoupled, so the neutrino temperature should be somewhat less than the 3 K radiation temperature; estimates suggest a temperature of about 2 K. Unfortunately, neutrinos interact very weakly with matter (whence the "weak" force that describes their interactions) and are therefore very difficult to detect: they can pass through light years of lead; every second about 100 *trillion* of them, most coming from nuclear reactions taking place in the core of the Sun (and some from the decay of radioactive potassium in bananas!), pass through your body—can you feel them?

As the temperature dropped below the threshold for electron-positron pair production, electrons and positrons annihilated each other without being replenished, leaving only the small excess of electrons due to the matter–antimatter asymmetry mentioned in the previous section. (The same process occurred earlier at $t = 0.01$ s for protons and neutrons and their antiparticles, all of which have rest mass energies some two thousand times greater than electrons). For both of these reasons, neutrons were thereafter destroyed faster than they were created and were no longer in thermal equilibrium with protons. The reactions

$$n \rightleftarrows p + e + \bar{v} \; (\beta\text{-decay; recall Note 10})$$
$$n + v \rightleftarrows p + e + 0.8\,\mathrm{MeV}$$
$$n + e^+ \rightleftarrows p + \bar{v} + 1.8\,\mathrm{MeV}$$

which at high temperatures proceed, as indicated, in both directions, then proceeded predominately to the right with the result that protons increasingly outnumbered neutrons with decreasing temperature. When $kT < 0.8$ MeV at temperatures just below 10^{10} K, the mean time for the middle reaction shown above exceeded the age of the Universe at that epoch (at about $t = 2$ s) and the neutron-to-proton ratio N_n/N_p froze out ("*neutron freezeout*") at a value of $e^{-1.29/0.8} = 1/5$. As we shall see in the next timestep, this ratio is critical in determining the abundance of the lightest elements, hydrogen and helium.

$$\text{Nucleosynthesis: } t = 3 \text{ min; } T = 10^9 \text{ K}$$

During the first three minutes in the history of the Universe (to borrow from the title of Physics Nobel laureate Steven Weinberg's account of the origin of the Universe [16]), all of the hydrogen and most of the helium— by far the most abundant elements in the Universe which is roughly three-quarters hydrogen and one-quarter helium by mass—formed from a cooling plasma of protons, neutrons, and electrons. Most of the neutrons became incorporated into helium nuclei—*alpha particles*[28]—through the following reactions:

$$n + p \rightarrow d + \gamma$$

$$p + d \rightarrow {}^3\text{He} + \gamma$$

$$d + d \rightarrow {}^3\text{He} + n$$

$$n + {}^3\text{He} \rightarrow {}^4\text{He} + \gamma$$

$$d + {}^3\text{He} \rightarrow {}^4\text{He} + p$$

Here, d denotes the *deuteron* (Greek *deuteros*, meaning "second"; second only to the proton, from the Greek *protos*, "first"), the simplest compound nucleus,

---

[28] In 1898, two years after the discovery of radioactivity by the French physicist Henri Becquerel (1852–1908), Ernest Rutherford, one of the greatest experimental physicists of all time, distinguished and named two types of particles emanating from radioactive materials: weakly penetrating α *rays* (*alpha rays*, shown by Rutherford in 1908 to be doubly ionized helium atoms, later recognized as bare helium nuclei known as *alpha particles*) and the more penetrating β *rays* (*beta rays*, electrons or their antiparticle twins, positrons, shown by Becquerel in 1900 to be the same as cathode rays). Rutherford and, two years later in 1900, the French physicist Paul Villard (1860–1934) independently discovered a third, even more penetrating emission, γ *rays* (*gamma rays*, so named in his 1904 book *Radioactivity* by Rutherford, who showed them to be penetrating electromagnetic waves of the shortest wavelength; recall Fig. 2.17). Unknown at the time of their discovery, these three distinct forms of radioactivity were labeled with the first three letters of the Greek alphabet, itself a word formed from its first two letters.

formed, as the first reaction shows, by the union of a neutron with a proton, and is thus an isotope of hydrogen, which can be written ^{2}H or simply D for the deuterium atom ("heavy hydrogen"), a deuteron encircled by an accompanying electron. (The two neutrons found in the heaviest hydrogen isotope, tritium, ^{3}H, from the Greek *tritos*, "third," makes it radioactively unstable and thus not important here.)

At temperatures above 10^9 K, deuterium is destroyed by the intense photon flux as fast as it's produced via the first reaction in the network shown above (i.e., the arrow points to the left instead of to the right as shown). As the Universe cooled, photodissociation of deuterium ceased and some neutrons β-decayed into a proton and an electron before being incorporated into a helium nucleus (recall the first of the three reactions listed in the previous timestep; the mean lifetime of a free neutron is about 15 min). Calculations indicate that due to this decay, the original neutron-to-proton ratio of 1/5 dropped to 1/7 [see, e.g., 21, p. 224]. Thus, for every 14 protons there were 2 neutrons, both of which quickly combined with 2 protons to give one helium nucleus for every 12 protons, giving a mass fraction for ^{4}He of 4/16 = 25% in excellent agreement with the observed cosmic abundance of helium. By number, hydrogen accounts for just over 90% of the baryonic matter in the Universe (Fig. 2.21).[29]

In the following few minutes, when the temperature dropped below 10^9 K as the Universe continued to expand and cool, the (positively) charged nuclei no longer had enough energy to overcome the mutual repulsion of the electric forces keeping them apart. This, together with the fact that the electrically neutral neutrons had all been incorporated into nuclei (mainly ^{4}He), means nuclear reactions essentially ceased and the strong force became unimportant in the early Universe. Apart from helium, a trace amount of ^{7}Li (lithium, a critical component in the lithium-ion batteries that power all our personal electronic devices, used also in the treatment of certain illnesses) was also synthesized. But because there are no stable nuclei having *atomic mass* (number of protons + neutrons) 5 and 8, and because the nuclear charge increases with *atomic number* (proton number, Z), no element heavier than lithium, already carrying three positively charged protons, was produced in any significant amount during primordial element building. As we shall see in the next chapter, we must look to the stars for the origin of the other elements, so many and so important to life, the Universe, and everything.

---

[29] Using a deck of cards to represent atoms in the Universe, in one deck of cards, the four aces would be helium atoms and the other 48 cards would be hydrogen atoms (approximately). In 30 decks of cards, you'd fine one carbon atom and a couple oxygen atoms, but you'd need 300 decks of cards to find one iron atom and over 38 million decks of cards to a gold atom.

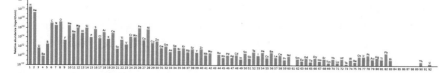

Fig. 2.21 The abundance by number of the chemical elements in the Solar System, essentially that of the Sun which contains 99.9% of the mass in the Solar System (with most of the rest coming from the two most massive planets, Jupiter and Saturn, both of which, like the Sun, are made mostly of hydrogen and helium), normalized to 10^6 for silicon; note the logarithmic scale. This pattern is ubiquitous in the Cosmos—and hence referred to as the "cosmic" abundance of the elements—indicating a common process of element synthesis throughout the Universe. The first thirty elements make up 99.99997% of the Sun. Slightly more than 99% of all the baryonic matter in the Universe is in the form of hydrogen and helium, with all but a very small fraction of the helium (which is produced in stars; see Fig. 3.14) synthesized during the first three minutes in the history of the Universe. The mass of all nuclei heavier than iron amounts to less than one-thousandth the mass of everything from lithium to iron. Elemental abundance would match the nuclear binding energy curve (see Fig. 3.10) had the elements been created in equilibrium conditions. (*Wikimedia Commons*, https://commons.wikimedia.org/wiki/User:Tkarcher, https://creativec ommons.org/licenses/by-sa/3.0/)

The agreement between predicted and observed values of the abundances of the two stable isotopes of hydrogen, ^{1}H and ^{2}H (10^{-5}), and those of helium, ^{3}He (10^{-5}) and ^{4}He, along with that of ^{7}Li (10^{-10}), the most abundant stable isotope of lithium, is, as David Arnett concludes, "a striking success for cosmological nucleosynthesis theory" [20, p. 143], indeed one of the greatest achievements of Big Bang theory. The abundances of the other naturally occurring isotope of lithium, ^{6}Li, and near neighbors in the periodic table, ^{9}Be (beryllium; ^{8}Be being extremely unstable to α-decay due to the extremely tight binding energy—and hence stability—of ^{4}He nuclei), a light metal used in aircraft construction, and the two stable isotopes of boron, ^{10}B and ^{11}B, found in the cleaning product borax, are produced over time by *cosmic ray spallation*, the breakup of heavier nuclei by collisions with high-energy cosmic rays, typically protons (see Fig. 3.14).

The stability of ^{4}He, the most stable of the light nuclei, makes its abundance relatively insensitive to density, but the deuterium abundance drops sharply with increasing density as it is so reactive; ^{3}He and ^{9}Li exhibit a smaller density dependence. Measurements of the abundances of these light elements indicate that between 1 and 5% of the Universe is made of normal baryonic matter, a result consistent with a completely independent estimate of between 4 and 5% from the fluctuation spectrum of CMB anisotropies (see the next timestep [21, p. 225]; recall that the rest of the "stuff" in the Universe is about 25% dark matter and 70% dark energy, neither of which we understand). Less than 10% of these baryons are in stars and galaxies; most are in a tenuous and pervasive hot intergalactic gas: most of the baryonic mass in the Universe is in the form of hot plasma, very different from the familiar matter around us here on Earth, and thus a bit of a blow to our long-held, if non-Copernican, expectations for every-day matter. And, in yet another example of how the physics of the very small (particle physics) and the physics of the very large (cosmology)—the microcosm-macrocosm connection (recall Note 22 and Figs. 1.2 and 2.1, and see Fig. 5.1 and Note 1 in Chap. 5)—"complement each other and enrich our understanding of the Universe as a whole," the cosmic helium abundance is consistent with experiments at CERN that limit the number of neutrino varieties to three, one each for the electron, the *muon* (a kind of "heavy" electron), and the *tau particle* (oxymoronically the most massive of the three leptons at nearly twice the proton mass) [22, p. 52]; the muon and tau particle are unstable, short-lived leptons that do not contribute in any significant manner to primordial nucleosynthesis.

$$\text{CMB:} t = 380{,}000\,\text{yr}; \; T = 3000\,\text{K}$$

With the decoupling of radiation and matter, the two were no longer in thermal equilibrium and the electromagnetic interaction became unimportant, leaving only gravity to shape the large-scale development of the Universe which was then electrically neutral. (If, for example, Earth and Sun each had an excess electrical charge—positive over negative or vice versa—of only one part in 10^{36}, the electrical—"Coulomb"—repulsion between them would be greater than their gravitational attraction; recall that like charges repel each other, opposites attract.) Matter, both normal baryonic matter and, even more importantly, dark matter, was then free to consolidate to form stars, galaxies, and galaxy clusters.

Because the mass-energy of a nucleon is about 939 MeV whereas the energy of a 3 K photon is about 0.0007 eV, even with a billion photons per nucleon most of the energy of the Universe today is in the form of matter. Radiation dominated before the temperature dropped below about 4000 K, just before matter and radiation decoupled some 380,000 years after the Big Bang when the temperature reached 3000 K (Fig. 2.16's "Afterglow Light Pattern"), because the energy density of baryons decreases with the third power of distance between them whereas the energy density of radiation falls as the fourth power of distance. As summarized by Weinberg [16, p. 76]:

> The enormous energy density of radiation in the early universe has been lost by the shift of photon wavelengths to the red as the universe expanded, leaving

the contamination of nuclear particles and electrons to grow into the stars and rocks and living beings of the present universe.

$$\text{Reionization:} \, t = 10^8 - 10^9 \, \text{yr}; \, T = 20 - 60 \, \text{K}$$

The first stars formed and their radiation reionized the Universe, forever ending the "Dark Ages" (recall Figs. 2.16 and 2.20). The Universe has reached puberty.

$$\text{Era of dark energy:} \, t = 8 \, \text{Gyr}; \, T = 5 \, \text{K}$$

The accelerated expansion of the Universe, the second since the initial inflation immediately following the Big Bang, is thought to have begun when the Universe entered its dark-energy-dominated era some 6 billion years ago (recall Fig. 2.16). Within the framework of general relativity, this accelerated expansion is described by a positive value of Einstein's "biggest blunder": the cosmological constant Λ. The cosmological constant has the effect equivalent to the presence of a positive vacuum energy, a kind of repulsive gravity dubbed "dark energy," driving the acceleration. For reasons we don't yet understand, empty space contains an inherent energy that causes it to expand.

$$\text{Today:} \, t = 13.8 \, \text{Gyr}; \, T = 2.7 \, \text{K}$$

Today, nearly 14 billion years after the Big Bang, we find ourselves bathed in a background of remnant radiation from the early Universe. Astronomers continue to study and learn more about this important message from "deep time" that even geologists would covet, now with very sensitive space-based instrumentation. The era of precision cosmology has arrived.

The first space mission specifically designed to study the CMB was the COsmic Background Explorer (COBE), launched by NASA in 1989. Among its key discoveries were that the CMB, averaged across the entire sky, conforms precisely to a thermal blackbody spectrum at a temperature of 2.73 K, and also shows very small temperature (and hence density) fluctuations on the order of 1 part in 100,000 across the sky, significant findings that earned the project coordinators John Mather and George Smoot the 2006 Nobel Prize in Physics. Without these small-scale inhomogeneities in the early Universe, galaxies and stars and planets and people would never have formed.

NASA's second generation CMB space mission, the Wilkinson Microwave Anisotropy Probe (WMAP) was launched in 2001 to study in more detail the very small fluctuations that were imprinted on the CMB at the moment the photons and matter decoupled 380,000 years after the Big Bang (Fig. 2.22). These fluctuations originated as sound waves ("baryon acoustic oscillations") at an earlier epoch immediately after the Big Bang and later grew under the influence of gravity, eventually giving rise to the large-scale structure (i.e. clusters and superclusters of galaxies) we see around us today. WMAP's results have helped determine the proportions of the fundamental constituents of the Universe—normal baryonic matter (5%), dark matter (25%), and dark energy (70%)—and established the Standard Model of cosmology on firm footing today.

The European Space Agency's Planck satellite was launched in 2009 to study the CMB in even greater detail than ever before. Covering a wider frequency range at higher sensitivity and smaller angular resolution than WMAP, it confirmed to an even higher degree the results of earlier measurements and refined our measure of the rate of expansion of the very early universe, complementing observations of supernovae which measure the expansion rate in the more recent past.

At the other end of the EM spectrum, the German-built X-ray satellite eROSITA (extended ROentgen Survey with an Imaging Telescope Array; the

Fig. 2.22 The detailed, all-sky picture of the infant Universe created by the Wilkinson Microwave Anisotropy Probe (WMAP) satellite. This "baby picture" of our Universe, capturing microwave radiation from 380,000 years after the Big Bang, is the equivalent of taking a picture of an 80-year-old person on the day of its birth. The image reveals 13.77 billion-year-old temperature fluctuations (shown as color differences, red for warmer and blue for cooler, with a range of ± 200 microkelvin) that correspond to the seeds that grew to become galaxies—and eventually, us. (*Photograph courtesy of NASA / WMAP Science Team, public domain*)

German physicist Wilhelm Röntgen discovered X-rays in 1895), has made the most detailed map of the X-ray sky and found some of the strongest evidence yet that dark matter shaped the cosmos. Probing nearly nine billion years of cosmic evolution by tracing the X-ray glow of distant clusters of galaxies, early results support the Standard Model of cosmology, according to which the gravitational pull of dark matter—a still-mysterious substance[30]— is the main factor shaping the Universe's structure: the lumpiness of the Universe is as lumpy (due to the tug of war between attractive gravity and repulsive dark energy) as the standard lambda-cold dark matter (Λ-CDM) cosmological model predicts (a model that inspired the 1991 mixed-media installation piece *Cold Dark Matter: An Exploded View* by the English artist Cornelia Parker). Furthermore, the galactic-cluster data enabled the team to tease out the role of neutrinos in shaping this cosmic web. As noted earlier, copious amounts of these elusive elementary particles were produced in the early Universe, and their low masses and reluctance to interact with other particles imply that they act like dark matter, forming haloes around galaxies. Analysis indicates that neutrinos could have masses of no more than 0.22 eV (an electron has a mass of 511,000 eV, more than a million times more), the tightest estimate of the neutrino mass to date (2024).

The quest continues to understand the Universe, past, present, and future, and most importantly, our place in it—our cosmic connection. One thing is certain: without the matter–antimatter asymmetry and the synthesis of hydrogen in the early Universe, we wouldn't be here. And if the Universe continues to expand, anyone still here will be the lonely hearts of the Cosmos.

* * *

Writing at the middle of the twentieth century, the British evolutionary biologist Julian Huxley noted that "the present is the first period in which we have been able to grasp that the Universe is a process in time" [23, p. 13]. Through science, we can now address the fundamental questions asked over a century ago by the artist Paul Gauguin's: *Where do we come from? What are we? Where are we going?* (recall Fig. 1.1). The American astronomer Robert Jastrow concluded his popular book, *God and the Astronomers* [24, p. 116], pointing to the analogy between modern scientific (Big Bang) and ancient religious (Genesis) ideas concerning the creation:

---

[30] As is the recently discovered "doesn't matter," which, unlike dark matter and antimatter appears to have no effect on the Universe whatsoever. And then there's the problem of how to deal with that "other" dark matter as we continue on our collision course with nature in the current Anthro-poo-cene. Dark times, indeed.

For the scientist who has lived by his faith in the power of reason, the story ends like a bad dream. He has scaled the mountains of ignorance; he is about to conquer the highest peak; as he pulls himself over the final rock, he is greeted by a band of theologians who have been sitting there for centuries.

George Smoot, one of the principal investigators of COBE, has dubbed the spectacular results of that mission, all in agreement with Big Bang cosmology and full of clues to the origin of the Universe, "the face of God"—as if it were God who actually lit the fuse. The words of Galileo, writing in his 1632 *Dialogue Concerning the Two Chief World Systems*, seem presciently relevant here, as all of this "might perhaps be fabled to have occurred in primordial chaos, where vague substances wandered confusedly in disorder" [25, p. 21]. To paraphrase the late Harvard evolutionary biologist Stephen Jay Gould, we live in one helluva fascinating Universe, whatever its modalities of construction.

The rapidly expanding early Universe, hot enough at the beginning, cooled within the first few minutes of its history after forming only hydrogen and helium in any substantial abundance, leaving the synthesis of the rest of the chemical elements, all called metals by astronomers (much to the dismay of chemists who know better—and without which we would not exist), to other places and times: *all metals are made in stars during their life and death*. As Eddington correctly surmised, hydrogen and helium, the ashes of Creation, would be the fuel for the stars. Primordial Big Bang nucleosynthesis set the stage for stellar nucleosynthesis, without which there would be no chemistry and hence no biology: helium is inert and you can't do much with hydrogen.

And so, *ad astra!* As Eddington replied to those who thought that the centers of stars are not hot enough to produce stellar energy (and chemical elements) by the nuclear fusion of four hydrogen atoms into helium, there is no "hotter place" science knows of—Eddington's "hotter place" being that spiritual realm of perpetual fire beneath Earth where the wicked are punished after death. Indeed, as Eddington reasoned, "What is possible in the Cavendish Laboratory may not be too difficult in the Sun" [26, p. 19]. As we shall see, without these stellar nuclear furnaces that give us life-sustaining light and heat during the day, and that we have for so long admired in our night sky, we simply wouldn't be here and the Universe would be a very different—and very lonely—place.

References

1 Y. N. Harari, *Sapiens: A Brief History of Humankind, "Part One: The Cognitive Revolution"* (HarperCollins Publishers, New York, 2015)

2 B. Malinowski, *Magic, Science and Religion and Other Essays* (The Free Press, Glencoe, IL, 1948)

3 E. Hadingham, *Early Man and the Cosmos* (University of Oklahoma Press, Norman, 1985)

4 B. Bobrick, *The Fated Sky: Astrology in History* (Simon & Schuster, New York, 2005)

5 R. Fleck, *Entropy and the Second Law of Thermodynamics ... or Why Things Tend to Go Wrong and Seem to Get Worse* (Springer, Cham, Switzerland, 2023)

6 P. W. Atkins, *The Second Law* (W. H. Freeman & Co., New York, 1994; orig. publ. 1984)

7 R. Tarnas, *The Passion of the Western Mind: Understanding the Ideas that Have Shaped Our World View* (Ballantine Books, New York, 1993)

8 J. D. Barrow, *Cosmic Imagery: Key Images in the History of Science* (W. W. Norton, New York, 2008)

9 T. S. Kuhn, *The Copernican Revolution: Planetary Astronomy in the Development of Western Thought* (Harvard University Press, Cambridge, MA, 1957)

10 S. Okasha, *Philosophy of Science: A Very Short Introduction*, 2nd edn (Oxford University Press, Oxford & New York, 2016; orig. publ., 2002)

11 L. M. Principe, *The Scientific Revolution: A Very Short Introduction* (Oxford University Press, Oxford & New York, 2011)

12 P. J. Bowler, I. R. Morus, *Making Modern Science: A Historical Survey* (University of Chicago Press, Chicago and London, 2005)

13 H. Kragh, *Quantum Generations: A History of Physics in the Twentieth Century* (Princeton University Press, Princeton, 1999)

14 A. D. Aczel, *God's Equation: Einstein, Relativity, and the Expanding Universe* (Four Wall Eight Windows, New York & London, 1999)

15 H. S. Kragh, D. Lambert, "The Context of Discovery: Lemaître and the Origin of the Primeval-Atom Hypothesis," Ann. Sci. **64**, 445–470 (2007) https://doi.org/10.1080/00033790701317692

16 S. Weinberg, *The First Three Minutes: A Modern View of the Origin of the Universe* (Basic Books, New York, 1977; rev. ed., 1993)

17 S. Toulmin, J. Goodfield, *The Discovery of Time* (Harper & Row, New York, 1965)

18 J. Gribbin, *The Scientists: A History of Science Told Through the Lives of Its Greatest Inventors* (Random House, New York, 2003)

19 P. M. Dauber, R. A. Muller, *The Three Big Bangs: Comet Crashes, Exploding Stars, and the Creation of the Universe* (Addison-Wesley Pub. Co., Boston, 1995)

20 D. Arnett, *Supernovae and Nucleosynthesis: An Investigation of the History of Matter from the Big Bang to the Present* (Princeton University Press, Princeton, 1996)

21 D. Maoz, *Astrophysics in a Nutshell* (Princeton University Press, Princeton & Oxford, 2007)

22 J. Barrow, *The Origin of the Universe* (Basic Books, New York, 1994)

23 J. Huxley, *Evolution: The Modern Synthesis* (George Allen & Unwin, Ltd., London, 1942)

24 R. Jastrow, *God and the Astronomers* (W. W. Norton & Co., Inc., New York & London, 1978)

25 G. Galilei, *Dialogue Concerning the Two Chief World Systems*, trans. Stillman Drake (Random House, New York, 2001)

26 A. S. Eddington, "The Internal Constitution of the Stars," Nature **106**(2653) (1920)

3

Stellar Evolution: The Life Cycle of Stars

Summary Stars, like people, have a life cycle. They are born, they age, and finally they die. But stars take a lot longer than people to go through their cycle of life: our Sun, for example, is a typical, middle-age star and yet is already nearly five billion years old. And just as we must deal with gravity in our everyday lives, gravity is the primary driver of stellar evolution, and therefore a star's mass is its most important property governing its structure and evolution. Stars are born when gravity compresses interstellar clouds of dust and gas to temperatures and densities high enough to sustain the thermonuclear fusion of hydrogen into helium, converting some of the original mass into energy in accordance with Einstein's famous equation, $E = mc^2$. Half the stars in the Universe, those having no more than half the mass of the Sun, end their lives as burned-out, Earth-size helium white dwarfs. The stronger gravity of more massive stars compresses their central regions to the higher temperatures required to fuse the heavier elements necessary for building the stuff of life before ejecting most of this "stardust," either quiescently as stellar winds and planetary nebulae or explosively as supernovae and kilonovae, into the surrounding interstellar medium—a star's loss but life's gain—leaving behind a stellar graveyard littered with white dwarfs, superdense neutron stars the size of a small city observed as pulsars, each having a mass comparable to the Sun, and black holes, the ultimate death star in which gravity is totally triumphant and so strong that not even light can escape them. The properties of each are discussed further in relation to our cosmic connection.

© The Author(s), under exclusive license to Springer Nature
Switzerland AG 2024
R. Fleck, *We Are Stardust*, https://doi.org/10.1007/978-3-031-67275-0_3

Twinkle, Twinkle, Little Star,
How I wonder what you are....

—Ann Taylor, "The Star" (*Rhymes for the Nursery*, 1806)

O Star... Tell Us What Elements You Blend.

—American poet Robert Frost, "Choose Something Like a Star" (1916)

We've come a long way in our understanding of the stars since Anaxagoras's fifth-century-BC suggestion that the stars are fiery stones and that the Sun—an average star, we now know—is a hot rock, bigger than Greece's Peloponnesian peninsula. Astronomy followed the advice of the sixth-century-AD commentator, Simplicius, who recommended that "the astronomer ... must go to the physicist for his first principles." In discussing the relation of physics to astronomy in his famous 3-volume *Feynman Lectures on Physics*, the Nobel Prize-winning American theoretical physicist Richard Feynman (1918–1988) had this to say about what he claimed was "the most remarkable discovery in all of astronomy" [1, p. 3–6, 3–7; emphasis in the original]:

> ... that *the stars are made of atoms of the same kind as those on the earth...* Strange as it may seem, we understand the distribution of matter in the interior of the sun far better than we understand the interior of the earth. What goes on *inside* a star is better understood than one might guess from the difficulty of having to look at a little dot of light through a telescope....

We've come a long way, too, since the French father of positivist philosophy Auguste Comte (1798–1857), believing the stars too distant to give up the secrets of their constitution, issued the following warning about the stars in his *Cours de Philosophie Positive* (1835; emphasis added):

> While we can conceive of the possibility of determining their shapes, their sizes, and their motions, *we shall never be able by any means to study their chemical composition or their mineralogical structure....* The innumerable stars that are scattered through space serve us scientifically only as providing positions.[1]

Philosophical prognostication regarding the limits of science must not be trusted (recall Cicero's quip in ancient Rome that "there is nothing so ridiculous but some philosopher has said it"), as the unreachable stars soon became objects of laboratory investigation through the new science of *spectroscopy*—the study of the series of colored bands and lines produced as light,

---

[1] Several years later, Britain's Astronomer Royal, George Biddell Airy, reiterated Comte's argument, warning that speculation about the nature of celestial bodies "possesses no proper astronomical interest."

more generally electromagnetic radiation, is spread out into its component features—a science that began with Isaac Newton's "very pleasing divertisse-ment" with a triangular glass prism that he used to spread sunlight into its component colors. Light—Feynman's "little dot of light," the *fiat lux* of Genesis—would become the starry messenger (to borrow from the title of Galileo's famous account of the many "more things in heaven and Earth, Horatio, than are dreamt of in your philosophy!"). In the nineteenth century, the astronomer became the astrophysicist, effecting a change in the discipline as profound as Galileo's use of the telescope initiated a new agenda of the natural philosophy of the heavens. This new science of spectroscopy readily disproved the French philosopher Auguste Comte's claim that the stars are too distant to give up the secrets of their true physical nature.

Comte was assuming that the determination of their chemical makeup would require physical samples of the stars—obviously, a very difficult propo-sition even today, even for the nearest star, our Sun. But in 1859 the German physicist Gustav Kirchhoff (1824–1887; famous also to students of physics for his laws governing electrical circuits) discovered that the chemical compo-sition—and more—of a gas could be deduced from the spectrum of its light viewed from an arbitrary distance. This powerful analytical technique was extended to astronomical bodies by the pioneering astro-spectroscopist William Huggins (1824–1910), who in 1864 England was the first to use a *spectroscope* attached to a telescope to study "the light of the stars and other celestial bodies." This precision instrument, used to produce and measure spectra, was, together with the camera, one of the two most important instru-ments introduced into science in the nineteenth century. Huggins declared (emphasis added) that the "One important object of this original spectro-scopic investigation … namely to discover whether *the same chemical elements as those of our earth are present throughout the Universe*, was most satisfac-torily settled in the affirmative; a common chemistry, it was shown, exists throughout the visible universe" [2, p. 49]. (The element helium, the second most abundant in the Universe, was first identified in the spectrum of the Sun, *hēlios* in Greek, some twenty-seven years before it was found on Earth.) By far the richest source of information about the Universe, spectroscopy is used to measure chemical abundances, temperature, density, pressure, motion, magnetic fields, and many other properties of objects near and far, some of which are over 10 billion light years away. The new science cast new light on the internal composition and workings of the stars, including, most importantly for our purposes here, the source of their energy and, as a conse-quence, the origin of the chemical elements, so essential to life, the Universe, and everything.

Joseph von Fraunhofer *Optiker und Physiker 1787-1826* Deutsche Bundespost
1987

Fig. 3.1 A German postage stamp celebrating the 200th anniversary of Joseph Fraunhofer's birth showing the solar spectrum drawn and colored by Fraunhofer with the dark lines named after him. (*Wikimedia Commons, public domain*)

The invention of the spectroscope by the Bavarian optician Joseph von Fraunhofer (1787–1826) and the subsequent development of the science of spectroscopy, pioneered by Kirchhoff, gave science its most powerful diagnostic tool for understanding the nature of both light and matter, both here on Earth and elsewhere in the Universe. Fraunhofer eventually identified, measured, and labeled 574 dark lines in the solar spectrum—"Fraunhofer lines"—assigning, for example, the letter D to a prominent line later linked to sodium (Fig. 3.1). He noticed similar lines in the spectrum of Venus, not surprising to us today since we now know that planets shine by reflected sunlight, and in the spectra of several bright stars. Kirchhoff noticed that some of Fraunhofer's dark lines correspond to some of the bright lines identified with the laboratory flame spectra of various elements such as sodium, *the first identification of the presence of any element beyond Earth.*[2] It was discovered that *different chemical substances display a different set of spectral lines with no two exactly alike, making the spectrum like a fingerprint—actually more like a UPC barcode* (Fig. 3.2)—*that can be used to identify the presence of chemical substances and to ascertain their properties, such as temperature, pressure, state of motion and the like, solely from the light they emit or absorb.*

Thus did the new science of spectroscopy reveal in the nineteenth century, contrary to the pessimism of Monsieur Comte, the constitution of the stars.

---

[2] Kirchhoff was awarded the Royal Society's Rumford Medal "for his researches on the fixed lines of the solar spectrum," which included a prize in gold sovereigns. "Look here," Kirchhoff told his banker, who cared nothing for unretrievable gold in the Sun, "I have succeeded at last in fetching some gold from the Sun."

Fig. 3.2 Emission line spectra of hydrogen (*top*) and iron (*bottom*). In stark contrast to the universal nature of the rainbow-like *continuous* thermal blackbody spectrum (also visible here; see also Fig. 3.9) explained by the German physicist Max Planck in 1900 (recall Chap. 2), the characteristic barcode-like *discrete* ("interrupted") line spectra of different gases was known since the early nineteenth century to be markedly different for different elements, each one exhibiting its own unique pattern of light at well-defined wavelengths. Here, wavelength increases from left to right across the visual range from 400 nm (blue) to 700 nm (red). The prominent red line in the spectrum of hydrogen is the H_α Balmer line produced when the electron in a hydrogen atom drops from the third to the second Bohr orbits, emitting light having a wavelength of 656.3 nm corresponding to the energy difference between the two quantum states, a transition responsible for the red coloration in so many emission nebulae such as the Orion Nebula (see Fig. 3.5). In 1885 the Swiss schoolteacher Johann Jakob Balmer (1825–1898), a numerologist trained in mathematics, noticed a numerical relation among the wavelengths of the first four hydrogen spectral lines (three of which, red, blue, and violet, are noticeable here) that in the early years of the twentieth century provided the key to interpreting the structure of the atom (see Note 3). (*Wikimedia Commons, public domain*)

Later, in the early twentieth century, spectroscopy would literally bring to light the structure and behavior of the atom, as attention became focused on the cause of spectral lines; in its laboratory setting, as yet another example of the connection across scales, macro to micro, it provided the experimental basis for quantum physics.[3] Physicist and acclaimed author Jim Al-Khalili proclaimed [3, p. 101] the "fact that we can learn about the ingredients of the Universe just by studying the light that reaches us from space" to be "one of the most beautiful notions in science." I agree: as an astronomy student,

---

[3] In 1913 the Danish physicist Niels Bohr (1885–1962), working with Ernest Rutherford in Manchester, combined Rutherford's nuclear ("solar system") model of the atom with Einstein's quantum model of light to give the first semi-quantum model of the atom, explaining, among other things, how atoms emit and absorb light at well-defined discrete frequencies (colors), an accomplishment that earned him the 1922 physics Nobel Prize. Bohr postulated that there exists a discrete set of allowed electron orbits, each of definite size and energy, and each different and unique to each chemical element, and that an atom emits or absorbs radiant energy discontinuously when one of its electrons "jumps" from one allowed orbit to another, with the accompanying change in the energy of the atom equal to the energy of the quantum of radiation that is emitted or absorbed in accordance with the Planck-Einstein equation $E = hf$, where h and f are the Planck constant and frequency of the radiation, respectively (as introduced in the previous chapter).

I remember being amazed at the (astronomical) amount of information that can be extracted from the little light we receive from objects so faint and so far away. I still am.

Born on the spine of spectroscopy at the end of the nineteenth century, theoretical astrophysics emerged in the first decades of the new century as a research program with the construction of mathematical models of stellar structure and evolution. The principal players were the pioneer astrophysicist introduced in the previous chapter, Sir Arthur Eddington, Cambridge University's 1904 "Senior Wrangler" and later Plumian Professor of Astronomy at Cambridge and director of the Cambridge Observatory, and the English mathematical physicist and astronomer Sir James Jeans (1877–1946), like Eddington, a Cambridge Wrangler. Their methodologies, like those of Plato and Aristotle, were complementary. Jeans, the disciplined mathematician, having spent his early career moving through a series of mathematics professorships, believed that the deductive certainty of pure mathematics was the only approach to understanding the Universe that could produce coherent, reliable knowledge. Eddington, on the other hand, a one-time chief assistant at the Royal Observatory with laboratory experience in Manchester, practiced a phenomenological pragmatism based on, and sometimes extending, the known laws of physics together with observation and experience.

Many of the discoveries of twentieth- and twenty-first-century astrophysics were made at the focus of giant new telescopes on the ground and later in space which provided an observing platform unimpeded by Earth's atmosphere (Fig. 3.3). By the end of the twentieth century, fueled largely by government support, astronomers were sampling the entire electromagnetic spectrum from gamma rays to radio waves (recall Fig. 2.17). The detection of neutrinos from the Sun in the 1960s and from an exploding star in 1987 (see Fig. 3.21), and the direct detection of gravitational waves in 2015, extended our astral reach into multi-messenger astronomy, bringing high-energy particle astrophysics and general relativity into the fold.

As discussed in the previous chapter, normal baryonic matter makes up only a little less than 5% of the Universe, and less than 10% of that is in stars. Thus, stars make up only about half a percent of the Universe. But thank God for stars. As we'll see in this chapter—and as I hope you'll appreciate when leaving—we wouldn't be here without them. *As a byproduct of energy generation, stellar processes produce the raw material for new generations of stars, for planets, and, at least in one instance, for ponies and people.*

Fig. 3.3 Kitt Peak National Observatory, located in the Arizona-Sonoran Desert on the Tohono O'odham Nation (from whom "the men with long eyes," as they call astronomers, lease the land), about 50 miles west-southwest of Tucson, Arizona. With more than twenty optical and two radio telescopes, it is one of the largest gatherings of astronomical instruments in Earth's northern hemisphere. Baboquivari Peak, just right of center on the horizon, is sacred to the O'odham Nation who revere it as the center of the Universe. (*Photograph by the author*)

3.1   Star Formation

We had the sky up there, all speckled with stars, and we used to lay on our backs and look up at them, and discuss about whether they was made or only just happened.
—Huck Finn, in Mark Twain's *The Adventures of Huckleberry Finn* (1884)

The first step to life in the Universe—after the Universe came to life—began with the formation of the first stars. When I was an astronomy graduate student in the mid-1970s, black holes—dead massive stars—were the singular hot topic of the day, strange beasts that helped generate a renewed interest in Einstein's general theory of relativity. The other end of the life cycle of stars, star formation, was of little interest to most people, including astronomers (although I have to believe that most everyone, me included, has some interest in origins—of just about anything and everything; recall Note 2 of Chap. 2). The discovery of *exoplanets*—planets orbiting other stars—at the end of the twentieth century changed that: star and planet formation are now trendy topics in astronomy. We now know that *planet formation is a natural byproduct of star formation. A good thing, that. We wouldn't be here if things were otherwise.*

One of the most easily recognizable constellations in the northern hemisphere winter sky is Orion, named for a hunter in Greek mythology and one of the 48 constellations listed by Ptolemy in late antiquity (Fig. 3.4). There's a lot here of relevance to the life cycle of stars. For example, the fuzzy reddish middle "star" in Orion's "sword" is the famous Orion Nebula (M42 in the Messier catalog of nebulous objects; Fig. 3.5a), one of the brightest nebulae in the sky, visible even to the naked eye, and, at a distance of just under 1500 light years, the closest region of massive-star formation to Earth (light reaching Earth tonight left it not long after the last Roman emperor left town). Some of the youngest stars known, having estimated ages of only a few hundred thousand years—stars live much longer than people[4]—are found here along with even younger, still-forming, *protostellar* (pre-stellar) objects (Fig. 3.5b).

The bright blue supergiant star Rigel, marking the left knee of Orion (*Rijl*, Arabic for "leg, foot"; *Rijl Jauzah al Yusrā*, "left leg of Orion"),[5] is the brightest and, at about 20 $M_\odot$ (twenty times the Sun's mass $M_\odot \simeq 2 \times 10^{30}$ kg; $\odot$ is the symbol for the Sun), the most massive star in a multiple star system of at least four stars (single stars like our Sun are rare). One of the largest stars known, its size (radius) is nearly 100 $R_\odot$ (100 times the Sun's radius $R_\odot \simeq 700{,}000$ km which in turn is 100 times that of Earth, meaning you could fit $100 \times 100 \times 100 =$ one million Earths inside the Sun) and its *luminosity*—its power output, the energy released per unit time expressed in watts, abbreviated W—is about 10^5 $L_\odot$ (100,000 times the Sun's luminosity $L_\odot \simeq 4 \times 10^{26}$ W, a *very* bright light bulb, indeed!). A star's energy reserve is determined by its mass through the Einstein mass-energy equivalence formula, $E = mc^2$, and so knowing a star's mass (and hence energy content) together with its luminosity (how fast it spends its energy), we can estimate how long it will take the star to go broke—to run out of energy (just like calculating how long it will take to go broke by dividing the amount of

---

[4] It's important to appreciate the enormous difference in scales—length, mass, and time, among others—at the human and cosmic levels (recall Note 12 of Chap. 2). As the American poet Robert Frost declared in his 1928 poem *On Looking up by Chance at the Constellation*, "You'll wait a long, long time for anything much / To happen in heaven...." Watching our Sun, which is halfway through its 10-billion-year life, age over an entire human lifetime would be like watching a human age for 10 s of time. Not much happening in either case.

[5] Astronomy was especially important in the Islamic religion for determining the start of each lunar month and the time and direction for daily prayer. The hot days and cool, clear nights characteristic of the region's desert climate provided further impetus for the study of astronomy; too hot for camel caravans during the day, desert merchants traveled at night, navigating by the stars. The names of the stars Aldebaran, Altair, and Betelgeuse, for example, are from Arabic names for the brightest stars in the constellations Taurus, Aquilla the eagle, and Orion, respectively. The prefix "al-" is Arabic for "the," and is used to make a noun definite. Other familiar "al-" words of Arabic origin, so common in the language of science and mathematics, are *al*gebra, *al*gorithm, *al*chemy, *al*cohol, and *al*kali.

Fig. 3.4 The constellation Orion (do you see the hunter?). The reddish nebula (*below center*; the middle fuzzy "star" in Orion's "sword" hanging down from the three bright stars of his "belt") is the stellar nursery known as the Orion Nebula (see Fig. 3.5; the red coloration is light emitted by hydrogen atoms). The yellowish star Betelgeuse (*left center*; Orion's northeast shoulder), is one of the brightest stars in the sky; Rigel is the bright white star to the right of the nebula marking the southwest knee of Orion. The yellowish star Aldebaran (*top center*) is the red eye of Taurus the Bull. In the course of the night, Orion chases Taurus westward across the sky, pursuing the Pleiades (see Fig. 3.6), the daughters of Atlas placed in the sky on the shoulder of the bull by Zeus for protection. (*Photograph by the author*)

money you have by the rate you spend it). Doing this for Rigel gives it a maximum age of about 8 million years: Rigel did not shine in the sky over the dinosaurs, all of which, with the exception of the avion varieties that remain with us today as birds, went extinct some 66 million years ago.

The red supergiant star Betelgeuse, located in the right shoulder of Orion, is, like Rigel, one of the brightest stars in the sky, has a similar mass and luminosity, and has a diameter nearly 1000 times that of the Sun: it would extend beyond Mars and the asteroid belt, nearly to Jupiter, if put in the place of our Sun, giving it a bulk density one hundred thousand times less than that of the air we breathe (far too low to generate enough lift on an airplane—in case you're planning a trip). It's nearing the end of its life and, at a distance of about 500 light years, could very well become the nearest supernova stellar explosion sometime in the next 100,000 years, at which time it will be nearly as bright as a full moon, easily visible in the daytime sky for several months. Stay tuned.

Fig. 3.5 a The Orion Nebula M42, a womb with a view. Measuring a couple dozen light years across, the Orion Nebula is a "blister" on the surface of a giant molecular cloud several hundred light years in size, spanning the entire Orion constellation. Barely discernable in the heart of the nebula, just to the *lower left of center*, is the Trapezium Cluster, named for its geometric shape formed by the four brightest members responsible for much of the illumination of the surrounding nebula, "placental" interstellar material—dust and gas, predominately hydrogen—left over from their formation less than half a million years ago. (*Photograph courtesy of NASA, ESA, M. Robberto (Space Telescope Science Institute/ESA) and the Hubble Space Telescope Orion Treasury Project Team, public domain*) **b** The Orion Nebula is home to many young and forming stars, nearly 200 of which are encircled with protoplanetary disks ("*proplds*"), fledgling planetary systems, several of which are shown here. (*Photograph courtesy of NASA, ESA, M. Robberto (Space Telescope Science Institute/ESA), the Hubble Space Telescope Orion Treasury Project Team and L. Ricci, ESO, public domain*)

Clearly, *star formation is an ongoing process in the Galaxy* which is nearly as old as the Universe itself and more than twice the age of our Sun and Solar System. It's an evolutionary scenario first suggested over two hundred years ago by the German-born English astronomer and freelance musician William Herschel (1738–1822), the discoverer in 1781 of the planet Uranus, the first planet to be discovered in historical times and an accidental byproduct of his comprehensive series of observations of the "fix't" stars. Herschel shifted the emphasis in astronomy from the ancient and traditional concerns with relatively local lights—the Sun, Moon, and planets—moving against the black background of the sky, to the mysteries of the background itself. Emphasizing the *temporal* rather than the *spatial* properties of the heavens, he suggested an evolutionary sequence with unresolved nebulae (like that in Orion; recall Fig. 3.5) being younger than star clusters (like the Pleiades and Omega Centauri; see Figs. 3.6 and 3.7). Here was *observational* support for the contemporary Laplacian *theory* of nebular evolution, evidence for a sequence of changes taking place over rather extended periods of time, a *historical development of the heavens* decoratively described by the Victorian Poet Laureate Alfred, Lord Tennyson (intended for his 1833 "The Palace of Art"):

> Regions of lucid matter taking forms,
> Brushes of life, hazy gleams,
> Clusters and beds of worlds, and bee-like swarms
> Of Suns and starry streams.

Herschel proposed that some "clustering power," a "condensing principle" he attributed to the "universal gravitation of matter," converted a diffuse nebula into a more condensed stellar body or cluster, so that the degree of "stellation" could be regarded as a sign of age, precisely as we today understand the processing of interstellar matter into stars. This new appreciation of evolution in the *physical* Universe would, in the following century, spill over into the *biological* world and eventually displace the prevailing paradigm of a static, clockwork Universe as the new principium of science.

In yet another example of theory matching observation—the sine qua non of science—our understanding of star formation, now aided by computer modeling of many of the complexities involved (including but not limited to the presence of magnetic fields and supersonic turbulence), has progressed alongside observational evidence for star formation once beyond our reach. We know that stars form by the collapse of cool *interstellar clouds*, denser-than-average regions of the tenuous *interstellar medium*, the dust—submicron agglomerates of silicates and carbonaceous material encased by icy mantles—and gas, predominately hydrogen, that fills the space between

Fig. 3.6 The Pleiades star cluster, an asterism that looks like a "little dipper." The Pleiades may be represented in 17,000-year-old cave art in Lascaux, France, depicted floating above the shoulder of a majestic aurochs bull, just as they appear in the sky above the constellation of Taurus the Bull. Today, many farming communities the world over still mark their agricultural year according to the visibility of the Pleiades. (*Photograph courtesy of NASA, public domain*)

Fig. 3.7 The globular cluster Omega Centauri, the largest-known globular cluster in the Milky Way containing nearly 10 million stars and a total mass of several million solar masses packed within a diameter of roughly 150 light years, so densely packed that were you living within it, you could read a book at night under the light from a sky full of bright stars. Even at its distance of some 17,000 light years, it's one of the few globular clusters visible to the naked eye. (*Photograph courtesy of KPNO, NOIRLab, NSF, AURA, Blythe Guvenen, public domain*)

("inter") the stars ("stellar"), most of which is hot atomic and ionized gas, all of which amounts to less than 10% of the Galaxy's stellar mass.

Over a century ago, Sir James Jeans estimated the critical mass an interstellar cloud must have to undergo gravitational contraction against internal pressure support, which Jeans assumed to be thermal gas pressure: today we know that supersonic turbulence—and hence suprathermal velocities—are important to the dynamics of the interstellar medium. Setting the internal kinetic energy $\frac{1}{2}MV^2$ of a uniform-density spherical cloud having a mass M and internal thermal and turbulent motions at a characteristic speed V, equal to its self-gravitational energy $3/5\ GM^2/R$, where $G = 6.67 \times 10^{-11}$ N m^2/kg^2 is the Newtonian gravitational constant (N, m, and kg designating, respectively, the metric units of newtons for force, meters for length, and kilograms for mass) and $R = (3\ M/4\pi mn)^{1/3}$ is the radius of a cloud composed of a gas having a per particle mass m ($\simeq 2.3\ m_H$ for a cosmic abundance of hydrogen in molecular form H_2 and helium; m_H being the mass of a hydrogen atom) and a uniform particle density n (note that mass density mn = mass/volume = $M/(4\pi R^3/3)$), gives the critical ("Jeans") mass in solar-mass units:

$$M/M_\odot \simeq 8000\left(V/\text{km s}^{-1}\right)^3/\left(n/\text{cm}^{-3}\right)^{1/2} (Jeans\ mass),$$

where, as indicated, V is in km/s and n is in particles/cm^3. (Again, don't worry if you can't follow the math, which I've done here only to show how we arrive at useful quantitative expressions in science.) For giant molecular clouds, the largest self-gravitating clouds in the interstellar medium, such as the one associated with the Orion Nebula (recall Fig. 3.5), $V \simeq 10$ km/s and $n \simeq 100$ cm^{-3}, giving a Jeans mass of about $8 \times 10^5\ M_\odot$, in agreement with observations. Instabilities within these clouds, typically hundreds of light years across, lead to hierarchical fragmentation forming cool ($T \simeq 10$ K), dense ($n \simeq 10^4$ cm^{-3}, high for the interstellar medium but compared to the 10^{19} particles/cm^3 in the air we breathe, much lower than the best laboratory vacuum) star-forming clumps supported by thermal gas pressure (with a thermal speed $V = (3kT/m)^{1/2} \simeq 500$ m/s at that temperature; recall that k is the Boltzmann constant and m is the particle mass) having a Jeans mass now of order $10\ M_\odot$, matching that of high-mass stars.

Many young stars are found grouped together in *open star clusters*, the end result of star formation within fragmenting interstellar clouds. The Pleiades (M45 in the Messier catalog; Fig. 3.6) is perhaps the most famous of these, being the most obvious cluster to the naked eye (and hence figuring prominently in the mythologies of cultures over time around the world [4]). At a distance of only about 440 light years, it's one of the nearest to Earth (light reaching Earth tonight left its member stars just before Galileo turned his telescope to the skies). Although only the brightest six or seven stars are visible to the naked eye (known thus as the Seven Sisters, daughters of Atlas placed on the shoulder of Taurus to protect them from amorous Orion), it's a particularly rich cluster containing over a thousand stars, all formed within the last 100 million years—old for people, but young for stars.

The Jeans mass for self-gravitating clumps in our Galaxy during its formation shortly after the Dark Ages some 13 billion years ago (recall Chap. 2), when the gas temperature was about 2000 K and the particle density about 10^3 cm^{-3} is approximately 10^5 $M_\odot$ [5, p. 175], the mass of large *globular clusters*—the largest and most massive type of star cluster, spheroidal conglomerations of old stars bound together by gravity containing up to many millions of member stars—some 200 of which are found in the halo and central bulge of our Galaxy (Fig. 3.7). Globular cluster stars are among the oldest stars in the Galaxy and have metallicities only a fraction of the 1.7% solar metallicity. (Recall that all elements other than hydrogen and helium are considered metals by astronomers.) Astronomers distinguish two *stellar populations*: these old, low-metallicity stars are classified as population II; younger stars like the Sun are population I stars (easy to remember: "We're #1"), with the very youngest highest metallicity stars referred to as extreme population I. This immediately suggests that *metallicity in the Galaxy has increased over time as successive generations of stars convert primordial hydrogen and helium from the Big Bang into heavier elements through various energy-producing nuclear reactions in a process known as stellar nucleosynthesis.*

Protostellar condensations ("*protostars*") contract under their own weight and heat up (as a compressed gas does) until, after about a million years or so, core temperatures and densities become high enough to sustain nuclear fusion: a star is born, a self-gravitating spherical assemblage of hot plasma powered by thermonuclear reactions deep within its interior, a giant fusion reactor that begins its life on the so-called "main sequence."[6]

---

[6] Nuclear energy can be released either by splitting heavy nuclei (*fission*) like uranium such as is done in nuclear power plants, or by fusing together light nuclei (*fusion*) like protons to make helium as is done in the stars (see Fig. 3.10).

3.2   Life on the Main Sequence

Twinkle, twinkle little star,
I don't wonder what you are,
For by spectroscopic ken
I know that you are hydrogen.
        —attributed to English meteorologist Lewis Fry Richardson (1881–1953)

... for we do find
Seeds of them by our fire, and gold in them;
And can produce the species of each metal
More perfect thence, than Nature doth....

                                                    —Ben Jonson, *The Alchemist* (1610)

Stellar spectroscopy, which formed the core of the new discipline of astrophysics, came of age when, in 1890, the first Henry Draper Catalogue of stellar spectra was published by the Harvard College Observatory, introducing an alphabetical spectral classification scheme known as the Harvard Classification similar to that used in astronomy today. Listing the spectral types and magnitudes (brightness) of over ten thousand stars, the work was started by the New York astronomer Henry Draper (1837–1882) and continued after his premature death with generous funds donated by his surviving widow. Under the guidance of Harvard College Observatory director E. C. Pickering (1846–1919), who employed large teams of upwards of fifteen women including Draper's niece, Antonia Maury—"Pickering's harem"—in a "factory style" often marked by sharply gendered divisions of labor, male astronomers "manned" the telescopes while the women "computers," who would work diligently without expectation of high salary or career advancement—and in any case were typically not regarded as having the intellectual capability for truly creative work—read and classified the spectra.[7]

Some of the women made significant individual contributions. Most notable were those of Annie Jump Cannon (1863–1941) who extended the Henry Draper (HD) catalog in the early years of the twentieth century to include 225,300 stars, each with its identifying HD number (Betelgeuse, for example, is HD 39801; recall Fig. 3.4), refining the spectral classification system that in its general form remains essentially the standard still today designated (for historical reasons) with the letters O, B, A, F, G, K,

---

[7] During the two World Wars, the U.S. Army hired women to calculate artillery-trajectory tables, and, as the book and film *Hidden Figures* portrayed, women computers employed by NASA were crucial to the success of U.S. space program. Most of the "scanners" who examined the millions of detector photographs taken in the early days of high-energy particle physics during the mid-twentieth century were "girl observers."

M. Using the new physics of the quantum in her 1925 Harvard Ph.D. thesis, recognized as among the most brilliant in astronomy, the British-born American astronomer Cecilia Payne (later Payne-Gaposchkin; 1900–1979) demonstrated that this spectral sequence is a temperature sequence, with O stars, having surface temperatures exceeding 30,000 K being the hottest, and M stars—three-quarters of the stars in the sky—the coolest at just below 3000 K (the Sun is a G-type star with a surface temperature just under 6000 K).[8] Significantly, she showed also that *stars are composed primarily of hydrogen and helium*, with only minor differences in composition as inferred from their spectra, a groundbreaking conclusion that established the dominant role of hydrogen in the Universe, findings initially rejected because they challenged the contemporary view that there were no significant elemental differences between the Sun and Earth.

Payne-Gaposchkin showed that with stellar spectra, what you see is *not* always what you get. Although all stars are basically balls of hot hydrogen and helium plasma, the spectral signature of helium, for example, is strong only in O and B stars which are hot enough to energize the strongly bound electrons in the helium atom; hydrogen lines are dominant in A stars which have just the right temperature to energize the hydrogen atom, not too hot to drive the single electron away and not too cool to leave the atom dormant; metals dominate the spectra of cooler stars (that of singly ionized calcium being particularly prominent in solar-type G stars); and molecular absorption bands such as those due to titanium oxide (TiO) dominate the spectra of M-type stars that alone are cool enough to allow molecules to form. Astronomers had previously mistakenly surmised that the sequence of spectral types was an evolutionary sequence, with stars passing from blue to red as they lost energy and cooled in the course of their lives, giving rise to the terms "*early*" and "*late*" spectral types still used today.

By the early twentieth century, astronomers had collected data for a large sample of stars. But raw data—the "facts"—by themselves, have no meaning: data must be organized and interpreted (notwithstanding Mr. Gradgrind shouting in Dickens's *Hard Times* for "Facts, sir; nothing but Facts!").[9] For stars, the *Hertzsprung-Russell* (HR) *diagram* (Fig. 3.8), is *the*

---

[8] Each spectral class is subdivided into numerical subclasses using the numbers 0 (hottest) through 9 (coolest; unlike "hot" topics today getting the biggest numbers up to a perfect "10"); our Sun is a G2 main-sequence star. The entire sequence has been greatly expanded to include stars that don't fit the original scheme, and include such classes as D for white dwarfs and C for stars with strong carbon lines. I always thought it was a shame that the spectral type A0 bright star Sirius, the "Dog Star" marking the nose of the "Big Dog" constellation Canis Major, isn't a K9 star.

[9] "The edifice of science requires not only material, but also a plan," proclaimed the nineteenth-century Russian chemist Dmitri Mendeleev, famous for his periodic table, the plan of the chemical

conceptual scheme that organizes the collected data and allows us to make important generalizations concerning the nature and evolution of stars based upon correlations among their observed properties—in this case, luminosity and surface temperature (or the related properties, magnitude and color; whence a "color-magnitude" diagram)—an organizational plan analogous to Mendeleev's periodic plan for the chemical elements.

Created independently in 1911 by the Danish engineer and amateur astronomer Ejnar Hertzsprung (1873–1967) and by the Princeton University astronomer Henry Norris Russell (1877–1957) in 1913, the HR diagram

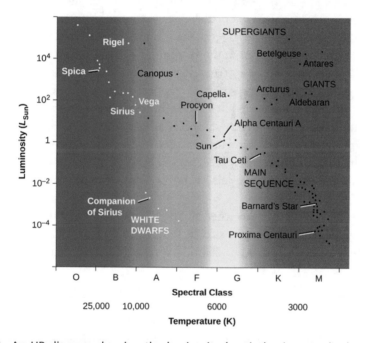

Fig. 3.8 An HR diagram showing the luminosity (*vertical axis;* note the logarithmic powers-of-ten scale) and surface temperature (and corresponding spectral class; *horizontal axis*) for some of the more familiar stars in the sky, including the bright red supergiant Betelgeuse, the bright blue supergiant Rigel, and the bright red giant Aldebaran, all shown in the sky in Fig. 3.4. Note that most stars fall along a diagonal "main sequence." Stars in the upper right are bright, cool, giants and supergiants, while those in the lower left are faint, hot, white dwarfs. The colored background indicates the temperature-dependent color of stars across the diagram. Keep in mind that the HR diagram is not a two-dimensional *spatial* map of stars as they appear in the sky (see Fig. 3.4 for that), but rather a *parameter* map of stellar luminosity and surface temperature. (*Photograph courtesy of Physics LibreTexts, CC BY 4.0 Deed*)

---

elements. Recall from the previous chapter that the German astronomer Johannes Kepler discovered the laws—the plan—governing planetary motion based on Tycho Brahe's observational data.

represents a major step towards our understanding of stellar structure and evolution. It was discovered that nearly 90% of the stars in the sky have luminosities and surface temperatures that cluster along a diagonal band in the HR diagram reaching from the upper left corner of the hot, bright O stars to the lower right corner of the cool, faint M stars, a band known as the *main sequence*. Just as with a large random sample of people, a large sample of stars will likely include stars of all ages, some just forming, some young, some middle-aged, some old, and even some dead, so the fact that about 90% of stars are main-sequence stars implies that *stars spend about 90% of their lives as main-sequence stars* and that these stars therefore maintain relatively constant luminosities and surface temperatures for most of their lives. Indeed, the main-sequence stage in the life of a star is both the longest and the most stable. And, importantly, in spite of Robert Frost's poetic warning in his 1928 *On Looking up by Chance at the Constellations* that "You'll wait a long, long time for anything much to happen in heaven" (recall Note 4), we can learn about the life cycle of the stars, despite the huge disparity in stellar and human lifetimes, simply by sampling a large number of them so that our sample will likely include stars at all stages in their life cycle—just as we can sample the human life cycle by looking at a large number of people of all ages rather than by following one person through an entire lifetime from birth to death.

To a good approximation, the radiation emitted by a star, like that of a blackbody, depends only on its surface temperature (Fig. 3.9). Thus, the well-established laws of blackbody radiation, such as Wien's displacement law,

$$\lambda_{\text{peak}}(\text{cm}) \simeq 0.3/T(\text{K}) \quad (Wien's\ displacement\ law),$$

introduced in the previous chapter (recall Note 20 of Chap. 2), can be applied to stars. According to Wien's law—and as can be seen in Fig. 3.9—the peak in the intensity of radiation from cooler stars occurs at longer wavelengths (mathematically, a smaller value for T in the denominator gives a bigger value for λ_{peak}), and hence these stars will appear redder in color than hotter stars which radiate more strongly at shorter (blue) wavelengths. A heated piece of metal exhibits the same color change from a dull red to a bright yellow to a brilliant bluish white as its temperature is increased. The bright star Betelgeuse in the shoulder of Orion, for example (recall Fig. 3.4), has a surface temperature of about 3600 K and thus a peak intensity at about 830 nm, just beyond the red end of the spectrum in the near-infrared and so looks red to the eye; the bright star Sirius, with a temperature of about 10,000 K, looks bluish white because its intensity peaks at about 300 nm, just beyond the blue end of the spectrum in the near-ultraviolet. Wien's law affords a

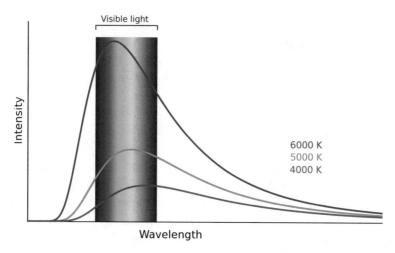

Fig. 3.9 The continuous spectrum of radiation emitted by a blackbody at three different temperatures. Clearly, more total radiation (the area under each temperature color-coded curve) is emitted at higher temperatures (Stefan's law), and more of that higher-temperature radiation is emitted at shorter (blue) wavelengths (Wien's law). (*Wikimedia Commons, public domain*)

simple and powerful way to determine the *temperature* of astronomical blackbodies—such as stars and, as discussed in the previous chapter, the CMB of the Universe: measure the wavelength corresponding to the peak intensity (i.e., the "color") and Wien's law gives the temperature. (After all, you can't stick a thermometer in a star's mouth to determine its temperature: stars don't have mouths—and, in any case, are much too far away.)

Figure 3.9 also illustrates another important characteristic of blackbody radiation, namely that the total amount of radiation over all wavelengths (luminosity L) increases dramatically with increasing temperature, a feature quantified by the *Stefan-Boltzmann equation*:

$$L = A\sigma T^4 = 4\pi R^2 \sigma T^4 \quad (Stefan-Boltzmann\ equation),$$

where a spherical star of radius R has a surface area $A = 4\pi R^2$, and $\sigma = 5.67 \times 10^{-8}$ W m^{-2}·K^{-4} is the Stefan-Boltzmann constant. (Again, please don't worry about the math, as we will continue to use very little of it and only when absolutely necessary to establish a quantitatively important result: it's how the various parameters relate to each other qualitatively, not quantitatively, that we really need to appreciate.) As one would expect, larger and/ or hotter bodies will radiate more total energy (which explains why more efficient radiators and cooling fins—and body parts of some animals—have larger surface areas allowing them to radiate more energy).

The Stefan-Boltzmann equation can be used to derive the *radius* of a star of known luminosity and surface temperature; nearly every other estimate of stellar radii is made empirically, mostly for stars that are observed to pass in front of each other as seen from Earth as the pair—known as *binary stars*, in this case, *eclipsing binaries*—orbit around a mutual center of gravity: knowing their orbital velocities, the eclipse duration provides a measure of the size of the star (short/long duration implying small/large stars).[10] Because $R \sim \sqrt{(L/T^2)}$, the brightest (large L) and coolest (small T) stars, those in the upper right corner of the HR diagram, are therefore the largest stars—the *supergiants* (bright, cool Betelgeuse, for example, is a red supergiant)—while the faint (low L) and hot (high T) stars in the lower left of the diagram are the smallest stars known as *white dwarfs* (white due to their high white-hot temperatures, dwarfs for their small sizes). Using $R \sim \sqrt{(L/T^2)}$ again, we see that for stars of similar luminosity (those occupying a narrow, horizontal band across the HR diagram), the cooler stars are the larger ones, and for stars of similar surface temperature (those having the same spectral type running vertically up the HR diagram), the brighter stars are the larger ones. Although stellar surface temperatures vary by only an order of magnitude (i.e., a factor of ten), stellar luminosities reach over ten orders of magnitude (powers of ten) from about 10^6 $L_\odot$ for the highest luminosity stars down to about 10^{-4} $L_\odot$ for stars with the lowest luminosities. Using $R \sim \sqrt{(L/T^2)}$, we find that stellar radii range from about one thousand times that of our Sun (1000 $R_\odot$) for the largest supergiants, nearly as large as the size of the orbit of Jupiter around the Sun, down to about one one-hundredth that of the Sun (1/100 $R_\odot$), comparable to the size of Earth.

Not much was known about the internal constitution of stars at the beginning of the twentieth century. Having a density comparable to that of water (1 g/cm^3) suggested a liquid structure—and significant compression by some twenty orders of magnitude to form from the much lower-density interstellar medium. Jeans, in particular, who had studied in detail the equilibrium shapes of rotating liquids, work still drawn on today, had long advocated liquid stars consisting primarily of high atomic number elements that would spontaneously decay into radiation and lighter elements. Eddington, writing

---

[10] Only our Sun is close enough to see it as a sphere and hence, knowing its distance and angular size, we can determine its radius (which simple math shows is equal to half its angular size expressed in radian arc measure times its distance from Earth). Some of the largest stars such as Betelgeuse have had their sizes estimated using an optical trick called *interferometry* which analyzes the interference of light waves coming from opposite sides of the star. *Lunar occultation* measurements have also been used to estimate stellar size by analyzing the light interference pattern produced just as a star disappears behind the limb of the Moon. General relativity, which predicts a size-dependent *gravitational redshift* of the light leaving a massive, compact object, has been used to estimate the size of some white dwarfs and neutron stars.

around 1910 in the *Encyclopaedia Britannica* entry for "Star," admitted that stars "might be solid, liquid, or a not too rare gas." But a few years later, Eddington generalized earlier models of the equilibrium of self-gravitating gaseous spheres to include radiation pressure—the pressure exerted by light— which he showed must play an increasingly important role in maintaining the equilibrium of stars of increasing mass. Focusing on how this radiation is transported within a star from its unknown source suspected to lie deep within a star, Eddington managed to reproduce many of the observed properties of stars, such as size, luminosity, and surface temperature, as well as interior pressure, temperature, and density, all while sidestepping *the greatest problem looming over stellar theory: the source of stellar energy.*

Eddington's crowning achievement, which became *the* guiding principle in stellar astrophysics early in the century, was his prediction, confirmed by observation, that the stellar luminosity of main-sequence stars increases sharply with stellar mass—the so-called *mass-luminosity law*—indicating that, despite having an average density comparable to water, stars can be described using the well-known ideal-gas laws (which explains why we know more about the interior of the Sun than we do about what lies beneath us here on Earth). Although the exponent varies slightly over different stellar mass ranges, the luminosity of main-sequence stars increases approximately as their mass cubed:

$$L \sim M^3 \quad (Mass-luminosity\ law).$$

For the range of luminosities for main-sequence stars, from 10^6 $L_\odot$ down to about 10^{-4} $L_\odot$, the mass-luminosity law predicts a corresponding range of stellar masses from 100 $M_\odot$ down to about 0.05 $M_\odot$, in agreement with the observed upper and lower limits for the mass of a star. In thermostat-like fashion, the intense radiation pressure from the photon flux flowing out through very hot luminous stars sets the upper mass limit with a corresponding maximum luminosity known as the *Eddington luminosity or Eddington limit* after its discoverer, and for these stars $L \sim M$. The inward force of gravity for masses less than about 0.07 $M_\odot$ is too weak to crunch their core densities and temperatures high enough to fuse hydrogen. These objects, ranging from about 80 down to 13 times the mass of Jupiter are "failed stars" known as *brown dwarfs*, and self-gravitating objects having a mass less than 13 times that of Jupiter are *planets* (more on planets, so important to life as we know it, in Chap. 4). Like a star's habitable zone (see Chap. 4), there is a Goldilocks-like range for stellar mass, not too hot and not too cold.

Eddington summarized his work on stellar structure in his 1926 ground-breaking monograph, *The Internal Constitution of the Stars*, which was immediately recognized as the authoritative work on the topic. The success of Eddington's models notwithstanding, it was soon realized that understanding the structure and evolution of stars required an understanding of stellar energy production.[11]

Early attempts in the nineteenth century to account for the energy produced by the Sun and other stars included the possibility of the energy released during chemical burning. It certainly seems as though the Sun is "on fire" and hydrogen seemed a likely fuel. The combustion of hydrogen (rapid oxidation: $2 H + O \rightarrow H_2O$) releases a specific (per unit mass) energy of about 10^8 J/kg, roughly three times the energy released when burning fossil fuels. Dividing this by the specific solar luminosity $\varepsilon_\odot = 4 \times 10^{26}$ W $\div$ 2 $\times 10^{30}$ kg $= 2 \times 10^{-4}$ W/kg, remembering that 1 W $=$ 1 J/s and that there are about 30 million seconds in a year, gives a chemical lifetime for the Sun of just 20,000 years, far less than the several hundred million years estimated for the age of Earth from the geological record at that time. (Today, based on radiometric dating of its rocks, we know Earth is several billion years old.) Clearly, chemical energy is not the source of stellar energy.

Another nineteenth-century proposal for the production of the Sun's energy was the gradual conversion of the gravitational energy driving its contraction during formation into thermal energy ("heat"), much as the gravitational potential energy of a falling stone is converted into its kinetic energy of motion. Physics (specifically, the *virial theorem*) shows that half of this gravitational energy is radiated away and half is stored as thermal energy E_{th} $= \frac{1}{2} GM_\odot^2/R_\odot$ [see, e.g., 6, pp. 48–49]. Dividing this energy by the solar luminosity gives the so-called *Kelvin–Helmholtz* (KH) *timescale*:

$$t_{KH} = \frac{1}{2}GM_\odot^2/R_\odot L_\odot \simeq 20 \text{ Myr} \quad (Kelvin-Helmholtz\ timescale),$$

a measure of how long the Sun could shine at its present luminosity with this energy source—and hence also an upper limit to the age of the Sun, which is what the British physicist Lord Kelvin (1824–1907) and the German polymath Heinrich von Helmholtz (1821–1894) wanted to know. Thus, without any other source of energy, the Sun would stop shinning about 20 million

---

[11] Eddington's 1920 address to the British Association, containing, according to one reviewer, "some of the most prescient statements in all of astronomical literature," including his prophetic predictions of the source of stellar energy, addressed also the prospect of harnessing the energy of the stars here on Earth: "If, indeed, the subatomic energy in the stars is being freely used to maintain their great furnaces, it seems to bring a little nearer to fulfillment our dream of controlling this latent power for the well-being of the human race—or for its suicide."

years after it formed, contrary to the geological and paleontological evidence that Earth—and hence the Solar System including the Sun, all of which formed at about the same time—is about 4.5 billion years old.[12] The KH timescale is a measure of the time it would take a hot, self-gravitating body like the Sun to lose its entire reserve of thermal energy at a steady luminosity, and is therefore a *thermal timescale*.

The KH timescale is also the time—the *pre-main-sequence timescale*—it takes slowly contracting pre-main-sequence stars-to-be to reach the main sequence and begin life as a star when core hydrogen fusion is initiated. Because these objects are large and cool, they make their first appearance on the HR diagram in the upper-right corner and slowly approach the main sequence along evolutionary tracks as their structure—and hence luminosity and surface temperature—change. Driven by gravitational contraction, the energy source of pre-main-sequence "stars," the most massive ones spend less than several tens of thousands of years in this pre-stellar stage, whereas the least massive can take several hundred million years to reach the main sequence, in all cases spending about one percent of their lives in this formative stage of their life cycle.

Other mechanisms suggested to account for the energy of the Sun and the stars, such as heat produced by constant meteoritic bombardment, were quickly dismissed as untenable. After the discovery of radioactivity in the late nineteenth century, many scientists thought that this process might provide the energy of the stars, which they accordingly assumed were comprised mainly of heavy elements, but, as we have noted, Cecilia Payne-Gaposchkin showed that the relative abundance of the elements in stellar atmospheres was nearly uniform from star to star with hydrogen being by far the most abundant element. At about the same time, Eddington and, independently, the young Danish theorist Bengt Strömgren (1908–1987), concluded that hydrogen must be the most abundant element not only *on* but also *in* stars in order account for the opacity—the degree of transparency—of stellar matter, a revolutionary recipe for the Universe resisted by most astronomers at the time.

In 1917 Eddington proposed the possibility of the annihilation of protons and electrons (later, after their discovery, positrons and electrons) as a viable and extremely long-lived source of stellar energy that could last for trillions of years. Three years later, based on the confirmed abundance of helium in stars and its rarity elsewhere, Eddington suggested as an alternative that

---

[12] Kelvin used this argument from physics, then the "king of the sciences," against Darwin, whose theory of natural selection required much more time to explain the diversity of life on Earth. Physics later proved Kelvin wrong.

the fusion of four hydrogen atoms into helium could be the source of stellar energy. But Eddington's suggestion fell on deaf ears: George Gamow's "quantum tunneling" proposal, such as allows an alpha particle to "tunnel" through the strong, attractive nuclear force during α-decay (discussed in the previous chapter), had not yet been worked out, and without "reverse tunneling"—penetrating the repulsive Coulomb barrier from the outside—the central temperature of the Sun, which Eddington estimated at 15 million K, was too low to overcome the electrostatic ("Coulomb") repulsion between protons (recall that like charges repel each other).[13] Eddington replied to his opponents that "we do not argue with the critic who urges that the stars are not hot enough for this process; we tell them to go and find a *hotter place*" (original emphasis)—in other words, they can go to hell! "What is possible in the Cavendish Laboratory may not be too difficult in the Sun," Eddington reasoned, referring to Rutherford's recent nuclear experiments. Indeed, discounting the possibility of hell, there are only two places hot and dense enough for nuclear fusion: the first few minutes in the history of the Universe (as we learned in the previous chapter) and deep within stars—not counting hydrogen ("H") bombs that release fusion energy explosively (also called "thermonuclear" bombs because of the high temperatures required to fuse hydrogen and its isotopes) and very recent controlled fusion laboratory experiments that may one day power the planet, a very hopeful and useful twist on "star power." The Big Bang was, in a sense, an explosive fusion

---

[13] At the 10^7 K temperature typical of stellar interiors, nuclei are moving with an average kinetic energy of order $kT \simeq 1$ keV, about 1000 times less than the MeV energies required to overcome the mutual Coulomb repulsion ($\sim q^2/r_N$, where $q = 1.602 \times 10^{-19}$ C is the elementary unit of electric charge and $r_N \simeq 10^{-15}$ m is the range of the nuclear force) between two protons, nuclei having the least positive charge and hence the least electrostatic repulsion. Even the (high-velocity) "Maxwellian tail") fraction of nuclei having MeV thermal energy is of order $e^{-E/kT} \simeq e^{-1000} \simeq 10^{-424}$, so that even with $M_\odot/m_H \simeq 10^{57}$ protons in the Sun, not a single nucleus—in the Sun or, for that matter, in *all* the stars in the observable Universe—has the energy required by *classical* physics to overcome the Coulomb barrier and undergo nuclear fusion with another nucleus. But just as a person without sufficient energy to climb over a wall might tunnel through it, the *quantum* wave nature of "particles" (well understood but mathematically beyond our scope here) allows nuclei to penetrate—to "tunnel" through—the Coulomb barrier. Because the probability of tunnelling is very low, fusion proceeds slowly and the nuclear fuel—and hence the star—lasts for "astronomically" long timescales. *Without quantum tunneling, stars would not shine and we would not be here.* Who said quantum physics doesn't matter much in everyday affairs? (Actually, our modern world is full of devices that operate on the principles of quantum physics, including the flash memory chips found in USB drives that use quantum tunneling to erase their memory cells, along with the semiconductors, transistors, and microchips that are the brains of all digital computing devices and cell phones, as well as lasers, such as those that scan your groceries at checkout and those that scan the digital "pits" of information coded into CDs and DVDs, the atomic clocks at the heart of GPS, magnetic resonance imagining (MRI) machines, which, in turn, rely on superconducting magnets, solar-panel photovoltaic cells and their quantum inverse—LEDs, light-emitting diodes, such as those used for high-efficiency lighting—and, coming soon, quantum computing, to name just a few.)

reactor just as the stars are controlled fusion reactors, nuclear pressure cookers cooking the ingredients for life.

Gamow's theory of α-decay, worked out in 1928, ushered in a new and less speculative chapter in *nuclear astrophysics* when it was realized that Gamow's inside-out quantum tunneling process might be reversed to explain the building up of the elements by nuclear reactions in stars. In that same year, highlighting the underlying macrocosm-microcosm connection between stars and atoms, Eddington reminded readers of his aptly titled *Stars and Atoms* that you must "look both ways. For the road to a knowledge of the stars leads through the atom; and important knowledge of the atom has been reached through the stars" [7, pp. 8–9]. But it was only after the discovery of the neutron in 1932 that nuclear astrophysics began to deliver promising results. Inspired by laboratory experiments, both Gamow and the British physicist Harold Walke (1912–1940) developed the neutron-capture theory of element formation and energy generation in stars. Walke pressed the analogy between laboratory and stellar processes [8, p. 184]:

> The atomic physicist, with his sources of high potential and his discharge-tubes, is synthesizing elements in the same way as is occurring in stellar interiors, and the processes observed, which result in the liberation of such large amounts of energy of the order of millions of volts indicate how the intense radiation is maintained and why their temperatures are so high.[14]

Nevertheless, these early ideas of neutron-induced nuclear processes did not satisfactorily explain either element formation in stars or stellar energy production. The breakthrough in solving the stellar energy problem came in 1938, and was very much a result of progress in nuclear theory and, to some degree, reflected the desire of some of the best, young, theoretical physicists to make their mark by doing work comparable to that of the creators of quantum mechanics a decade earlier.

The details were worked out independently by the German physicists Hans Bethe (1906–2005), a man who before tackling the problem of stellar energy generation "knew nothing about the interior of stars but everything about the interior of the nucleus," and in rather less detail by Carl von Weizsäcker (1912–2007), who worked with the quantum physics pioneer Werner Heisenberg in Leipzig. The essence of Bethe's theory [9], which became the foundation of all subsequent work—Russell hailed it as "the most

---

[14] Walke should have taken more care with these "sources of high potential": he died from the effects of an electrical shock received while working in the Liverpool University physics lab.

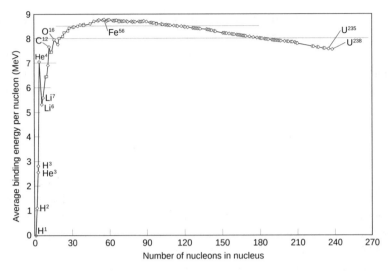

Fig. 3.10 A graph of the average binding energy per nucleon as a function of the number of nucleons (= protons + neutrons) in the atomic nucleus, also known as the *mass number* (*A*). Note that this nuclear binding energy, a measure of the energy required to remove a nucleon from an atomic nucleus—how deep "in the hole" a nucleon is—is, at roughly 8 MeV per nucleon for most elements, a million times more than the (ionization) energy that binds electrons in an atom. The unusual stability of ^{4}He, ^{12}C, and ^{16}O, which along with ^{1}H are the most abundant elements in the Universe, arises from an inherently stable nuclear shell structure analogous to that of the atom that accounts for the chemical stability of elements (with the inert gas ^{4}He being particularly stable at both the nuclear and atomic levels). The most stable nucleus—the one with the greatest binding energy per nucleon and thus most tightly bound together by the nuclear force—is that of iron-56 (^{56}Fe): heavier nuclei experience an increasing electrostatic repulsion from member protons (ripping apart nuclei heavier than uranium, none of which are stable and found in nature), while lighter nuclei present too much exposed nucleon surface, thereby diminishing the attractive nuclear force holding a nucleus together. "No other substance is so rigidly held together by the intertanglement of its elemental atoms as cold iron, that stubborn and benumbing metal," the Roman poet Lucretius perspicaciously declared in his first-century BC *De rerum natura* (*On the Nature of Things*). Nuclear energy is released by "climbing" the binding energy curve, either by splitting heavy nuclei (*fission*) such as the uranium shown here, or by fusing together light nuclei (*fusion*) like protons (^{1}H) to make helium (^{4}He) as is done in the stars. A graph of the abundance of the chemical elements (recall Fig. 2.21) would match the nuclear binding energy curve had the elements been created in equilibrium conditions. (*Wikimedia Commons, public domain*)

notable achievement of theoretical astrophysics of the last fifteen years"— was the *nuclear fusion of protons into helium nuclei with the release of nuclear binding energy* (Fig. 3.10) producing 4×10^{26} watts of power as 600 million (metric) tons of hydrogen is converted into 596 million tons of helium *every second*, with the difference in mass—4 million tons (a crash diet for you or me)—appearing as energy in accordance with Einstein's famous mass-energy equivalence equation, $E = mc^2$.[15] Bethe was awarded the 1967 Physics Nobel Prize "for his contributions to the theory of nuclear reactions, especially his discoveries concerning the energy production in stars," the first ever in astrophysics, by which time astronomy was finally formally recognized by the Nobel committee as a branch of physics. Von Weizsäcker, the son of a former German president, later became a member of an elite team of German physicists working on an unsuccessful attempt to develop a nuclear weapon for Nazi Germany during World War II.

A nuclear source for stellar energy explains Eddington's theoretical mass-luminosity relationship, as the stronger gravity of the more massive stars compresses the energy generation region in the stellar core to higher temperatures and densities, speeding up the nuclear reactions resulting in higher energy generation rates and hence higher luminosities. Just like calculating how long it will take to go broke by dividing the amount of money you have by the rate you spend it, it's easy to show that the Sun will shine for about 10 billion years, a time equal to its nuclear energy reserves (about one-tenth of $4/600$ of its mass times c^2) divided by the rate it spends it (i.e., its luminosity, 4×10^{26} W) and thus called its *nuclear* or *main-sequence timescale*. Stars live much longer than people and don't change perceptibly over a human lifetime, as Robert Frost reminded us (recall, again, Note 4). Since, according to Eddington's mass-luminosity law, stellar luminosity increases approximately as the stellar mass cubed ($L \sim M^3$), the main-sequence lifetime of a star ($\sim$ energy/luminosity $\sim Mc^2/L$) $\sim 1/M^2$, so that more massive stars—the high-luminosity big spenders that, like the American actor James Dean, live

---

[15] While 4×10^{26} W is certainly a lot of energy per unit time—dividing this number by the world's annual energy budget of about 5×10^{21} J shows you could power the world for nearly a *million years* on the energy that leaks from the Sun in just *one second*!—the human body is, pound for pound, much more luminous than the Sun. Dividing the solar luminosity (4×10^{26} W) by the mass of the Sun (2×10^{30} kg) gives a specific (per unit mass) solar luminosity $\varepsilon_\odot = 2 \times 10^{-4}$ W/kg. A typical 2000 Calorie per day food intake provides a person having a mass of 100 kg (equivalent to a weight on Earth of about 220 pounds) with a specific power of about 1 W/kg (= 2000 Cal/day $\times$ 1 day/86400 s $\times$ 4186 J/Cal $\div$ 100 kg), some 5000 times greater than that of the Sun. Stars are much brighter than people only because they are much more massive. Although nuclear energy sounds powerful—and indeed, with nuclear binding energy measured in MeVs (as shown in Fig. 3.10), it is a million times more so than chemical energy, which is measured merely in eVs (the nuclear bombs dropped on Japan during WWII each released the chemical energy equivalent of about 20 kilotons of TNT, each converting into that energy just *one gram* of matter [= E/c^2 = 20 kilotons $\times$ 4×10^{12} J/kiloton $\div$ $(3 \times 10^8$ m/s$)^2$])—it's not a very *efficient* source of energy: every second the Sun converts "only" 4 million out of 600 million tons of mass into energy, leaving behind—"wasting"—596 million tons of mass, for a mass-energy conversion efficiency of only $4/600$ = 0.7%. During matter–antimatter annihilation, *all* the mass is converted into energy with 100% efficiency, making this the most efficient energy generation mechanism in the Universe. As we'll see later in the chapter, even the strongest gravitational fields in the Universe associated with black holes are only 50% efficient in converting gravitational energy into other forms of energy.

fast and die young—exhaust their energy supply and burn out quicker than less massive stars. The most massive stars, which have a mass approaching $100\ M_\odot$, exhaust their nuclear fuel in just a million years, while the least massive stars having a mass just under a tenth of a solar mass will shine for a trillion (10^{12}) years. The bright star Rigel in the knee of Orion, for example (recall Fig. 3.4), will last only a few tens of millions of years and therefore did not light the night sky when the dinosaurs roamed Earth more than 66 million years ago. The difference in stellar lifetimes can be appreciated by following a spiral galaxy's changing color from the yellowish light of old—and hence low-mass—stars in the center to the young blue star clusters and reddish star-forming regions that typically delineate the outer spiral arms (recall Fig. 2.11a for M31, but other galaxies, such as the face-on NGC 1232, show the effect more dramatically; old, yellowish stars dominate elliptical galaxies and globular clusters, most of which have long ago exhausted the raw material to form new stars).

The nuclear fusion of hydrogen into helium—*hydrogen "burning"*—in stars takes place through two primary reaction networks: the proton-proton chain and, in stars slightly more massive than the Sun, the CNO cycle, the first stellar thermonuclear reaction network studied in the founding papers of the subject. (Note that "burning" is used figuratively here as a nod to the very high temperatures, typically measured in tens of millions of degrees, required for nuclear fusion, often labeled *thermo*nuclear fusion for that reason. It's much too hot inside stars for chemical burning or for any chemistry at all.)

The three reactions in the main branch of the *proton-proton ("p-p") chain* (with energy released and reaction timescale in parentheses) are [6, p. 49; see, also, Fig. 3.11]:

$$p + p \rightarrow d + e^+ + \nu \ (0.425\,\text{MeV, 10 billion years})$$
$$p + d \rightarrow {}^3\text{He} + \gamma \ (5.49\,\text{MeV, 1 sec})$$
$${}^3\text{He} + {}^3\text{He} \rightarrow {}^4\text{He} + p + p \ (12.86\,\text{MeV, 300,000 years}).$$

There are three different branches of the *p-p* chain using slightly different reactions but the three reactions shown here, comprising the PPI branch, are responsible for nearly 85% of *p-p* helium and stellar energy production; the other two branches, one of which (PPIII) is important for producing the Sun's highest energy neutrinos, include participation by lithium, beryllium, and boron, none of which survive. The positrons (e^+) quickly annihilate with electrons which remain stripped free from atomic nuclei at these high temperatures, producing two 0.511-MeV γ-rays. The neutrinos (ν), interacting so very weakly with matter, carry an average energy of 0.26 MeV directly out

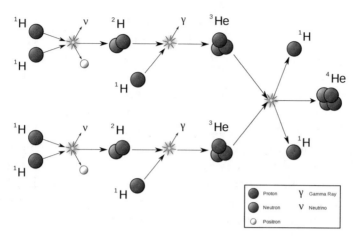

Fig. 3.11 The main branch (PPI) of the proton-proton (*p-p*) cycle converting four protons (¹H, *entering from the left*) into one helium nucleus (⁴He, *exiting to the right*). (*Wikimedia Commons, public domain*)

of the Sun providing our only direct window on stellar interiors via neutrino detectors here on Earth. (The mean free path of solar neutrinos is ten million times the solar radius; they can pass unimpeded through light years of lead.) The remaining energy carried by the photons and the helium nuclei is thermalized via mutual scatterings and collisions. The first reaction is very slow because it proceeds via the weak interaction (as evident by the emission of a neutrino) which converts a proton into a neutron that then combines with a proton to form a deuteron (*d*). Because this is by far the slowest of the three, it acts as a bottleneck and determines the timescale for the entire *p-p* chain— which is a good thing: otherwise, stars would have long ago burned through all the hydrogen in the Universe which would then be a very different place today.

Because two ³He nuclei are required in the third reaction to produce one ⁴He nucleus, the first two reactions must occur twice. Deducting the 2 × 0.511 MeV rest-mass energy of the two preexisting electrons, the total energy released for each ⁴He nucleus produced is therefore

$$(2 \times 0.511 + 2 \times 0.425 + 2 \times 5.49 + 12.86) \text{ MeV} = 25.71 \text{ MeV},$$

which (check for yourself!) equals the rest-mass difference between four protons and a slightly lighter ⁴He nucleus, the net result of the *p-p* cycle: $4\,p \rightarrow$ ⁴He. (Note that this energy can be estimated from Fig. 3.10 by adding up the differences in binding energies per nucleon between reactants and products.) Dividing this energy by the rest-mass energy of four protons (4×938.272 MeV) gives a mass-to-energy conversion efficiency of 0.7% (a number I always thought James Bond 0.007 would appreciate).[16] Taking the time derivative of Einstein's mass-energy formula—essentially dividing both sides of $E = mc^2$ by time—and setting the energy loss per time equal to the solar luminosity $L_\odot$, gives the rate of mass-to-energy conversion necessary to keep the Sun shining: a whopping *4 million tons per second* ($= L_\odot/c^2$). Dividing this mass-loss rate by the 0.007 conversion efficiency shows that 600 million tons of hydrogen are converted into 596 million tons of helium every second (numbers previously stated, now demonstrated—because science should be based on fact, not faith).[17]

The *nuclear timescale*, sometimes called the *main-sequence lifetime*, of a star producing energy via the *p-p* chain is therefore

$$t_{MS} \approx 0.1 \times 0.007 \times Mc^2/L \ (Main-sequence\ lifetime),$$

where the factor 0.1 reflects the fact that *a star's structure is uniquely determined by its mass and chemical composition*, a principle known as the *Vogt-Russel theorem* after the two astronomers who independently formulated it nearly a century ago: computer models confirm that when about 10% of a main-sequence star's hydrogen has been converted to helium, depleting essentially all the hydrogen in its hydrogen-burning core, its structure will

---

[16] Comparing the energy released during nuclear fusion to that released during nuclear fission, the fission of one uranium (^{235}U) atom releases an energy equal to about 200 MeV, nearly ten times that for a typical fusion reaction, but the uranium atom is more than 200 times heavier than hydrogen, so its mass-to-energy efficiency is about ten times less than that for fusion. Hence the push to develop the hydrogen bomb in the years following WWII. "So quick and ingenious are the minds of men in learning the most pernicious arts," lamented the 14th-century Italian humanist, Petrarch, when shown the new terror weapon of the time—the cannon—words that ring true even today.

[17] And it's nice in science to arrive at the same result from different directions (if you care to follow a little math). For example, although the timescales given in parentheses for the three reactions in the *p-p* chain are calculated from the details of nuclear physics on the microscale (e.g., quantum-mechanical reaction cross sections and reaction rates), the 10-billion-year timescale for the proton-to-deuteron reaction can be estimated from a consideration of the overarching macroscale. The number of reactions per second is given by dividing the solar luminosity (4×10^{26} W) by the reaction energy (25.71 MeV) to give 10^{38} reactions/sec (remembering that 1 W = 1 J/s and 1 J = 1 N m, where 1 N = 1 kg·m/s^2, and 1 MeV = 1.602×10^{-13} J). (Twice that number of neutrinos are produced every second, about 100 trillion of which pass through your body which stops about 100 of those every year. No wonder they're called "weakly interacting.") The total number of protons in the Sun is its mass (2×10^{30} kg) divided by the mass per proton (1.67×10^{-27} kg) $\simeq 10^{57}$ protons, of which only about 10% can participate before the composition will have changed enough, due to the conversion of hydrogen to helium, that the Sun will change its structure and start to evolve off the main sequence as its outer layers cool and expand to become a red giant star, thus ending its main-sequence lifetime. Dividing the 10^{38} reactions/sec by 10^{56} participating protons gives 10^{-18} reactions/sec, or, inverting, a wait time between reactions of 10^{18} s = 30 billion years, matching very closely the reaction timescale predicted by nuclear (micro)physics. Pretty cool—for such a hot environment.

have changed enough to move it off the main sequence (recall Note 17). Substituting in the solar values $M_\odot$ and $L_\odot$ gives a *main-sequence lifetime for the Sun of 10 billion years*—a million times longer than that for chemical energy, reflecting the million times more binding energy in the nucleus (MeV) compared to that in the atom (eV)—so at its current age of about 5 billion years, our Sun is a middle-age star halfway through its main-sequence lifetime. We've got plenty of time to find another home.

Stars today have about 1% of their mass in the form of carbon, nitrogen, and oxygen (CNO), but only those main-sequence stars more massive than about 1.5 $M_\odot$ (amounting to less than a few percent of the stars in the sky) have central temperatures high enough to produce most of their energy converting hydrogen to helium via the *CNO cycle* which uses trace amounts of CNO as catalysts to facilitate the reactions. Whereas less than 2% of the energy produced by the Sun comes from the CNO cycle, Sirius, the brightest star in the night sky, having a mass twice that of the Sun, produces most of its energy via the CNO cycle. Higher core temperatures (about 20 million K compared to 15 million K for the Sun; a minimum of 10 million K is required for the *p-p* chain) are necessary to overcome the greater electrostatic repulsion arising from a greater number of protons in the nuclei of CNO. For temperatures between 10 and 20 million K, the *p-p* energy generation rate scales as T^4 whereas that for the CNO cycle is much more temperature sensitive ($\sim T^{20}$), so that although core temperature varies little along the main sequence, the slightly higher temperatures in the more massive stars make the CNO cycle the dominant hydrogen-burning pathway.

The main branch of the CNO cycle as originally proposed by von Weizsäcker and Bethe in the late 1930s is illustrated in Fig. 3.12. Other branches involve other isotopes of CNO along with those of fluorine, all acting as catalytic intermediaries in steady state that do not accumulate in the star, and all have the same net result as in the *p-p* cycle: the conversion of four protons into one ^{4}He nucleus with nearly identical energies and efficiencies. *Significantly*, because it has the slowest proton-capture reaction rate in the entire cycle, *nitrogen (^{14}N), an essential element in amino and nucleic acids, the building blocks of proteins and the carriers of genetic information coded in DNA and RNA, is the dominant CNO nucleus remaining after CNO hydrogen burning.*[18]

---

[18] In his account of our cosmic connection [10, p. 94], author John Gribbin emphasizes that

...it is the [CNO] carbon cycle operating inside stars that produces the nitrogen on which life as we know it depends. It isn't just that the elements in your body have been manufactured inside stars and scattered through space in spectacular explosions—the nitrogen nuclei in your

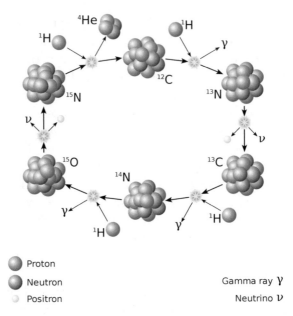

Fig. 3.12 The main branch of the CNO cycle as originally proposed by von Weizsäcker and Bethe in the late 1930s. Note that C, N, and O serve only as catalysts during the conversion of four protons (^{1}H) into one helium nucleus (^{4}He). (*Wikimedia Commons, public domain*)

After a long and stable stage converting roughly 10% of its hydrogen to helium, a main-sequence star undergoes significant structural changes as a result of this compositional change. The star then evolves off the main sequence (our Sun has already increased its luminosity by some 25% since it formed) and enters a *post-main-sequence stage* on the way to the stellar graveyard. As we will continue to appreciate, *a star's* mass *is the single most important parameter governing its structure and evolution*. It is to these post-main-sequence stages, populated by the likes of giants and supergiants and assorted other weirdos, many of which continue chemical element synthesis to heavier and heavier species, that we now turn our attention—main-sequence stars, after all, in synthesizing helium, give the Universe nothing new.

---

body, in particular, have been instrumental in determining the rate at which the carbon cycle proceeded in previous generations of stars.... [T]he very same nuclei that are now part of your body were once the dominant component of carbon cycle reactions going on inside stars. There is a direct connection between atoms in your body and the way in which stars at least one and a half times as massive as the Sun shine.

3.3   Star Death, Star Dust

Now we know that you went bust
Filled the void with clouds of dust.
Oxygen and carbon are
Elements made in a star.
Twinkle, twinkle, little star.
What you've made is what we are.

—Neal McBurnett

If the radiance of a thousand suns were to burst into the sky, that would be the
splendor of the Mighty One.

—*Bhagavad Gita* (second half of the first millennium BC)

You've heard about the Death Star, the moon-size space station with the ability to destroy an entire planet, featured in the film *Star Wars*. Well, now it's time to learn about star death.

Figure 3.13 illustrates the stellar life cycle—stellar evolution—from birth to death. Notice that *the mass of a star determines its life cycle and ultimate fate*, with most stars, those having less than about 8 $M_\odot$ at birth, peacefully passing away as slowly cooling white dwarfs, while more massive stars go out with a bang, ending up as either neutron stars or black holes. Many stars, regardless of mass, manufacture and disperse, one way or another, many of the chemical elements so essential to life, the Universe, and everything (Fig. 3.14). It's been said that a picture is worth a thousand words[19]; Figs. 3.13 and 3.14, which together present a pictorial précis of stellar evolution and the origin of the elements, are worth a million or more—if you care at all about your cosmic connection.

Already by 1957, in what is generally regarded as one of the most important papers in twentieth-century physics and astronomy [11], a comprehensive scheme for stellar nucleosynthesis (their paper was the first to use the term "nucleosynthesis") detailing the synthesis of most of the elements heavier than hydrogen and helium deep inside stars—a kind of stellar alchemy and a by-product of stellar energy production—had been worked out by the British-born astronomer Margaret (1919–2020) and physicist husband Geoffrey (1925–2010) Burbidge, together with the British cosmologist and

---

[19] Inasmuch as we are visual creatures, perhaps becoming even more so in today's video world, I agree with rocker Rod Stewart that "every picture tells a story"—an updated version of the *New York Evening Journal's* editor Arthur Brisbane's 1911 dictum that a picture is "worth a thousand words," all stemming from the Roman poet Horace's *Ut pictura poesis*.

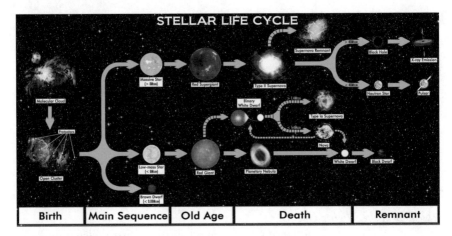

Fig. 3.13 The stellar life cycle from birth to death depends on the *mass* of the star, with low-mass stars, by far the majority of stars, going gently "into that good night," while massive stars "burn and rave at close of day; rage, rage against the dying of the light." (*Wikimedia Commons, R. N Bailey*, https://creativecommons.org/licenses/by/4.0/)

astrophysicist (later, Sir) Fred Hoyle (1915–2001; Fig. 3.15) and the American physicist William Fowler (1911–1995). As the saying goes, hydrogen and helium come from the Big Bang; Burbidge, Burbidge, Fowler and Hoyle (commonly written as the acronym B[2]FH) made all the rest.[20] Although some of the details remain uncertain—reaction rates (highly sensitive to temperature, whence "thermonuclear"), energy released, and the particular elements and isotopes produced by various astrophysical processes—the physics of stellar nucleosynthesis, one of the better-understood subjects of modern astronomy, is today modelled on computers and checked by multi-wavelength telescopic observations of stars in various stages of their life cycle

---

[20] Fortuitously, as predicted by Hoyle and later confirmed in the laboratory by Fowler, the energy of the very unstable beryllium-8 nucleus, formed by the combination of two helium-4 nuclei (alpha particles), when fused with a third alpha particle matches the energy of an excited state of the carbon-12 nucleus, a resonance that ensures that a small percentage of reactions produce carbon rather than decaying back to the original alpha particles. Hoyle realized that *if this resonance didn't exist, neither would we—or anything else in the Universe made of elements heavier than those cooked in the early Universe.* This fine tuning of the Universe to the "just so" conditions necessary for life, based as it is on carbon, is referred to as the *anthropic principle*, which is really nothing more than a simple logical requirement: intelligent beings cannot find themselves in a Universe uninhabitable by intelligent beings. (This starts to make more sense late at night after several glasses of wine.) Many other fine tunings that make our Universe possible have been identified, such as the value of the elementary unit of electric charge carried by the electron and the proton, and the value of many of the physical constants such as the Newtonian gravitational constant *G* which sets the strength of the gravitational force. Although surely self-evident, the reader is encouraged to ponder the cosmic significance—for life, the Universe, and everything (borrowing again from the *Hitchhiker's Guide* series)—of these finely tuned cosmic coincidences (again, late at night after plenty of wine).

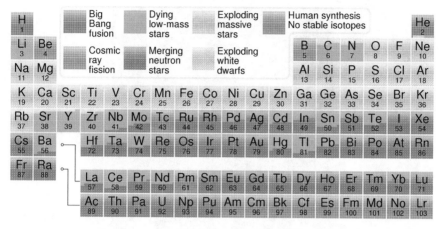

Fig. 3.14 A periodic table showing the probable origins of the chemical elements in the Solar System, prepared in early 2017 based on data by Jennifer Johnson at Ohio State University. The percentages of each element's origins are represented by squares (out of a hundred) to make it easier to estimate proportion. With few exceptions, the stars have given us the elements—"stardust"—so essential to life, the Universe, and everything, those exceptions being, of course, hydrogen and most of the helium which were formed shortly after the Big Bang (recall Chap. 2), and beryllium (Be), boron (B), and a small percentage of lithium (Li) which are produced over time by *cosmic ray spallation*, the breakup of heavier nuclei by collisions with high-energy cosmic rays, typically protons, and, finally, the synthetically produced unstable elements technetium (Tc) and promethium (Pm) and all elements heavier than plutonium (Pu). The origin of all but these artificially produced elements involve nucleosynthetic processes in the various color-coded astrophysical sites indicated here. Elements beyond lawrencium (Lw) are not shown. (*Wikimedia Commons*, https://commons.wikimedia.org/wiki/User:Cmglee, https://creativecommons.org/licenses/by-sa/3.0/)

and, when possible, tested by laboratory experiments (often difficult, ironically, because the energies of the reacting nuclei are so *low*—keV for reaction temperatures of order 10^7 K rather than the MeV, GeV, and TeV energies reached in particle accelerators), demonstrating agreement between theory, observation, and experiment that is the essence of science.

Because a star's fate is determined primarily by its mass, it's best to discuss the final stages in the life cycle of a star separately for the following three main-sequence mass ranges, each of which represents very different post-main-sequence and final stages (indicated parenthetically):

$$M < 0.5\,M_\odot \quad \text{(LOW-MASS STARS: no He burning; He WD)}$$

$$0.5\,M_\odot < M < 8\,M_\odot \quad \left(\begin{array}{l}\text{INTERMEDIATE-MASS STARS:}\\ \text{He burning; RG; PN; WD}\end{array}\right)$$

$$M > 8\,M_\odot \quad \left(\begin{array}{l}\text{HIGH-MASS STARS: advanced burning;}\\ \text{SG; WD(?); SN; NS; BH}\end{array}\right),$$

Fig. 3.15 Climbing a tower to the stars, astronomer Fred Hoyle, personifying *Pursuit*, is featured in the mosaic series *Modern Virtues* on the floor of the north vestibule of London's National Gallery, a mosaic record of the intellectual life of the 1930s and 1940s by the Russian mosaicist active in Britain, Boris Anrep, this one made in 1952. In the same series, *Curiosity* features Lord Rutherford with a splitting atom. (*Photograph by the author*)

where

WD   white dwarf
RG   red giant
PN   planetary nebula
SG   supergiant
SN   supernova
NS   neutron star
BH   black hole.

The story of these late and final evolutionary stages is summarized here in its essentials only, focusing particularly on the synthesis of the elements; the devil is always in the details—as the hundreds of research papers published every year on the subject confirm.

$$M < 0.5\,M_\odot$$

Stars having less than half the mass of the Sun—half of all stars[21]—go gently to the stellar graveyard after having slowly converted their hydrogen into helium. The main-sequence lifetime of a 0.5 $M_\odot$ star is nearly 100 billion years, much longer than the age of the Universe, which means that more than half of all the stars that have ever formed are still on the main-sequence. Theoretical modelling shows that most stars in this mass range, called *red dwarfs* or *M dwarfs* after their color or spectral type, remain fully convective throughout their lives, prolonging the period of hydrogen burning as hydrogen is continually dragged down to the hot core and converted into helium, eventually exhausting the star's entire supply of nuclear fuel.

These burned-out stars slowly contract, gradually increasing temperature, luminosity, and density, until electrons become packed so tightly together that *electron degeneracy*, a consequence of the *Pauli exclusion principle*—which limits the number of electrons to one per quantum state (otherwise all the electrons in all the chemical elements would occupy the same lowest-energy orbital and there would be no chemistry)—halts the contraction. Thermal gas pressure, even at these high temperatures, is not enough to support the star against gravitational collapse. The end result is a hot, helium *white dwarf*—a small, dense, Earth-size, burned-out star supported in the end against gravitational collapse by electron degeneracy pressure as electrons are forced into higher and higher energy states as lower-energy levels fill up (much as if skyscraper office space were filled from the ground floor up), exerting a considerable pressure *independent of temperature*, the same effect responsible for the hardness of metals. This temperature independence has the important effect of decoupling the mechanical structure of degenerate stars like white dwarfs from their thermal properties.

---

[21] About 90% of all stars in the solar neighborhood are less than the Sun's mass; most of the rest are less than twice the mass of the Sun, and only about 0.5% of all nearby stars are more massive than eight times the mass of the Sun.

The *Fermi energy* E_F, the maximum energy of a *fermion*,[22] such as the electron, the proton, and the neutron, in a completely degenerate ($T = 0$) gas, increases with density and is equal to about 100 keV at a density of 10^6 g/cm^3, the bulk density of a typical white dwarf. (Note the usage of the word degenerate here as a quantum effect, not a moral principle.) The corresponding *Fermi temperature* $T_F = E_F/k \simeq 10^9$ K, much higher than the thermal temperature of a white dwarf, reflects the much-higher-than-thermal energy of the star's degenerate electrons. The same is true for room-temperature (300 K) metals having a density typically of order 10 g/cm^3: their fermi temperature $T_F \simeq 10^6$ K, much higher than room temperature.

Degeneracy can also be understood in terms of the *Heisenberg uncertainty principle*, $\Delta p \Delta x \geq \hbar$ (where $\hbar$ is the reduced Planck constant: the Planck constant h divided by 2π), which states that there will always be some uncertainty, some spread in values (represented by the Greek letter Δ) in simultaneous measurements of momentum p and position x, small perhaps, because of the very small value of Planck's constant ($h = 6.626 \times 10^{-34}$ J·s), but not zero as would be the case in classical physics (for which h is nonexistent and would be set equal to zero here). Thus, as density increases, the electrons in a white dwarf are forced closer together into a smaller region of space Δx, so their spread in momentum (and hence pressure) becomes large because the product $\Delta p \Delta x$ is always greater than or equal to a constant ($\hbar$).

In 1931 the young Indian physicist Subrahmanyan Chandrasekhar (1910–1995; nephew of the Nobel Prize-winning physicist C. V. Raman) showed that the Fermi energy of a white dwarf is so high that its electrons move with speeds approaching that of light, which means that Einstein's theory of special relativity, which deals with effects of high-speed motion, must be incorporated into our understanding of these stars. Reworking the theory of the equilibrium structure of a white dwarf including relativistic effects, Chandrasekhar found that there is a maximum mass of about 1.4 $M_\odot$ that a stable white dwarf can have, a value now called the *Chandrasekhar mass*

---

[22] The term honors the Italian-American Nobel physicist Enrico Fermi (1901–1954) who developed the first statistical formulas governing particles that follow the Pauli exclusion principle, today known as Fermi–Dirac statistics in recognition of similar work done by the English Nobel theoretical physicist Paul Dirac (1902–1984), a Cambridge University Lucasian Professor of Mathematics (as was Isaac Newton and, in more modern times, Stephen Hawking) and one of the founders of quantum mechanics (he shared his Nobel Prize with Erwin Schrödinger, another quantum pioneer). One of the very few physicists to excel in both theoretical and experimental physics, Fermi is renowned especially for demonstrating the first self-sustaining nuclear chain reaction—the world's first nuclear reactor—at the University of Chicago as part of the Manhattan Project to develop the atomic bomb during World War II. Fermilab, the Fermi National Accelerator Laboratory outside Chicago, the U.S. version of Europe's CERN, is named after him.

or *Chandrasekhar limit*, M_{Ch}. The Chandrasekhar limit is the mass above which electron degeneracy pressure is insufficient to support a white dwarf—or, more generally, any electron-degenerate, self-gravitating mass, such as the core of a more massive post-main-sequence star (as discussed below)—against gravitational collapse.[23]

We'll return to white dwarfs in the next section which samples the stellar graveyard more broadly and in more detail. Besides, for our expressed purpose here to understand and appreciate our cosmic connection, with the exception of white dwarfs orbiting post-main-sequence close stellar companions[24] (discussed in the next section), more massive stars are more important for nucleosynthesis because they produce and disperse heavy elements on much more rapid timescales. Half the stars in the Universe never produce anything heavier than helium, all of which they take to the stellar graveyard.

$$0.5\,M_\odot < M < 8\,M_\odot$$

Stars having more than half the mass of the Sun never become fully convective and therefore develop a helium core surrounded by a hydrogen-burning shell and a hydrogen-rich envelope as they age. Lacking a source of energy, the inert helium core slowly contracts and heats up (as compressed gases do), increasing the temperature and hence the nuclear reaction rates in the surrounding hydrogen-burning shell. As a result of this increase in energy production and hence luminosity, the outer layers of the star expand and cool (as expanding gases do). Now with a higher luminosity and a lower surface temperature, the star evolves off the main sequence and begins its post-main-sequence stage, moving to the upper right of the HR diagram. It is now a *red*

---

[23] Chandrasekhar (the name means "he who carries the Moon") later became a U.S. citizen and spent most of his professional life at the University of Chicago. He was awarded half of the 1983 Nobel Prize in Physics for his "theoretical studies of the physical processes of importance to the structure and evolution of the stars" (the other half went to William Fowler, the "F" of B²FH fame). Chandrasekhar's still influential monographs, overflowing with mathematical elegance, span the fields of physics and astronomy: stellar structure (1939), stellar dynamics (1942), radiative transfer (1950), plasma physics (1960), hydrodynamic and hydromagnetic stability (1961), ellipsoidal figures of equilibrium (1969), and the mathematical theory of black holes (1983). NASA's third of its four "Great Observatories," the Chandra X-ray Observatory, is named after him.

[24] Roughly half of solar-type stars are formed as members of multiple star systems, and the multiplicity increases for higher stellar masses. Single stars like the Sun are relatively rare. A good thing that. Planetary motion controlled primarily by one, large, central mass (the Sun) has led some to label it—and its microcosmic equivalent, the hydrogen atom—a "gift of nature." It is entirely possible that, had we belonged to a multiple-star system (like Luke Skywalker's planet Tatooine in *Star Wars*) with its attendant complicated planetary dynamics, including the possibility of planets doing a "do-si-do" as they switch stellar partners (wouldn't that be a challenge for its inhabitants to figure out, assuming they survived the abrupt and severe changes sure to accompany such a wild dance of the planets!), we would still be asking ourselves "What makes the planets go?" and hence most likely not enjoying any of the uncountable products of modern science and technology at our disposal today.

giant star, a phase that lasts roughly one-tenth the main-sequence lifetime, from a billion years for solar-mass stars to only a few hundred thousand years for the most massive stars.

As the red-giant phase progresses, the core eventually reaches the 100 million K temperature required for helium burning through the two-step *triple-alpha* process:

$$^{4}\text{He} + {}^{4}\text{He} \rightleftharpoons {}^{8}\text{Be}$$
$$^{8}\text{Be} + {}^{4}\text{He} \rightarrow {}^{12}\text{C} + \gamma,$$

altogether releasing 7.3 MeV of energy. In the first step, unstable beryllium-8 decays back to two alpha particles (helium nuclei) in about 10^{-16} s. Nevertheless, a small equilibrium abundance of beryllium is established (only one ^{8}Be for every two billion ^{4}He at 10^{8} K) and, only because there exists an excited nuclear energy level in ^{12}C that matches almost exactly the energy of ^{8}Be + ^{4}He does the second step occur at all (recall Note 20 and see [12, pp. 119–126] for details). Recall from Chap. 2 that, due to the absence of stable nuclei with atomic masses 5 and 8, essentially no chemical elements heavier than helium were produced during primordial nucleosynthesis in the early Universe. This gap is bridged in hot, dense stellar cores by what is essentially a three-body resonance reaction: the triple-alpha process that converts three alpha particles into a carbon-12 nucleus, making "carbon-based units" such as us possible (to borrow a memorable phrase from *Star Trek: The Motion Picture*).

Helium burning is much more temperature sensitive ($\sim T^{40}$) than either one of the two hydrogen-burning networks, rendering the post-main-sequence phase much more susceptible to instabilities, such as brightness fluctuations and physical pulsations, than is the stable main-sequence stage. Its mass-to-energy conversion efficiency, found by dividing 7.3 MeV by the rest-mass energy of three alpha particles ($3 \times 4 \times 938.272$ MeV), gives a conversion efficiency just under 0.07%, a factor of ten smaller than that for hydrogen burning, which explains why the post-main-sequence lifetime is about one-tenth the main-sequence lifetime.

Once carbon is created in the core of a red giant, oxygen can be produced by the radiative *alpha-capture* reaction

$$^{4}\text{He} + {}^{12}\text{C} \rightarrow {}^{16}\text{O} + \gamma,$$

and, in turn, some of the oxygen can capture alpha particles to produce neon (Ne, as found in colorful electric lighting here on Earh), the next most abundant element in the Universe after CNO:

$$^4He + ^{16}O \rightarrow ^{20}Ne + \gamma.$$

Alpha-capture elements—those with even numbers of protons—are, not surprisingly, more abundant. In the reaction $^4He + ^{12}C \rightarrow ^{16}O$, there is an excited state of oxygen which, if it were slightly higher, would provide a resonance and speed up the reaction converting nearly all the carbon produced by the triple-alpha process into oxygen, leaving the Universe with very little carbon, so essential to life as we know it. And if an excited state of neon had an opposite parity (a quantum-mechanical symmetry property), most of the oxygen would be converted into neon. *Thus do we owe our existence as carbon-based units to the presence and absence of nuclear resonances involving alpha capture during stellar nucleosynthesis*: "small changes in seemingly boring excited states of nuclei could easily have led to a solar system in which boredom would not be a problem, because nobody could be around to be bored" [12, p. 128].

The increasing nuclear charge and hence increasing Coulomb barrier with increasing atomic number (proton number, Z) precludes synthesis of heavier elements, ^{24}Mg, ^{28}Si, etc., by radiative alpha-capture during helium burning. Oxygen and carbon are the main "ashes" of helium burning, and stars in this mass range, including our Sun which will never become hot enough for more advanced nuclear burning stages, eventually become carbon–oxygen (CO) white dwarfs. But not before shedding enough mass to end up with less than the Chandrasekhar mass limit of 1.4 $M_\odot$.

Indeed, due to the combination of low gravity in their extended outer regions (recall $g \sim 1/R^2$) and strong radiation pressure accompanying their high luminosity, red giants undergo large mass outflows—*stellar winds*—at rates up to $10^{-4} M_\odot$/yr, some of which is driven by recurrent thermal pulses of enhanced helium-shell burning beneath the shell of hydrogen burning, all of which slowly move outward though the star, leaving behind a hot core of carbon and oxygen. Often the result is a *planetary nebula* (Fig. 3.16; a misnomer because they are unrelated to planets but appeared planet-like when viewed through early telescopes), an expanding, glowing shell of gas containing an appreciable fraction of a solar mass stretching about one light year across, ejected at speeds ranging from 10 to 30 km/s. With an estimated 20,000 planetary nebulae in our Galaxy at any given time (each lasting only about 20,000 years, a very short time in the stellar life cycle), nearly 3000 of which have been catalogued and studied, they play an important role in the chemical evolution of the Galaxy, expelling a substantial amount of metal-enriched stellar material into the surrounding interstellar medium.

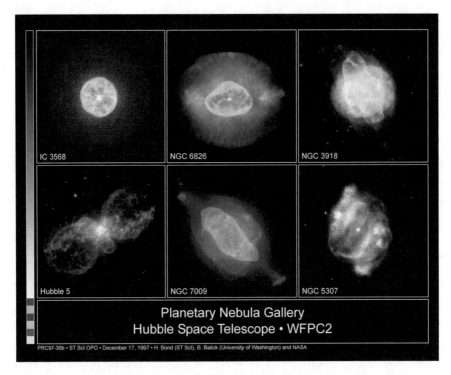

IC 3568   NGC 6826   NGC 3918

Hubble 5   NGC 7009   NGC 5307

Planetary Nebula Gallery
Hubble Space Telescope • WFPC2

PRC97-36b • ST ScI OPO • December 17, 1997 • H. Bond (ST ScI), B. Balick (University of Washington) and NASA

Fig. 3.16 A small gallery of various planetary nebulae taken by the Hubble Space Telescope, a tiny fraction of the many thousands known in the Galaxy. The varying and often complex morphologies of planetary nebulae are thought to be due to magnetic fields, the presence of a companion star, episodic ejections, or varying viewing angles from Earth. (*Photograph courtesy of H. Bond (ST ScI), B. Balick (University of Washington) and NASA/ESA, public domain*)

Although higher stellar masses are required for heavier-element nuclear fusion, *red giants in this mass range are the main contributors to the production of carbon and nitrogen, and they synthesize a significant amount of elements up to bismuth* (Bi), the most massive stable nucleus (recall Fig. 3.14), when seed nuclei successively capture neutrons, which, being electrically neutral, are not repelled by the electrically charged nuclei; α-decay limits the formation of heavier elements. *Neutron capture* comes in two flavors: the *slow process* (*s-process*) and the *rapid process* (*r-process*), the former taking place gradually in the low-neutron-density environment of red giants; the latter occurring rapidly in high-neutron flux environments such as exist during explosive stellar events like supernovae and neutron star mergers (see below). The timescales, slow and rapid, are relative to the half-life for the β-decay of neutron-rich nuclei (recall Note 10 of Chap. 2): if the β-decay half-life is short (long) compared to the timescale for neutron capture, the neutron-capture reaction is the s-process (r-process); during the r-process, neutrons are captured so quickly that the new, neutron-rich nuclei don't have time to β-decay. Free neutrons are produced during alpha capture by various isotopes (such as $^{13}C + {}^4He \rightarrow {}^{16}O + n$).

Direct evidence for s-process neutron capture is provided by the spectroscopic detection in red giants of the short-lived radioactive element technetium (Tc) that was first discovered in 1937 in debris from particle accelerators. (Technetium, a shiny gray metal, is named after the Greek word τεχνητός, "artificial," from the Greek *tekhnē*, "art" or "craft," the root of our word "technology," because it was the first element to be artificially produced.) Because red giants are much older than the longest Tc half-life of 4.2 million years, its presence in their outer atmospheres, brought up from deeper layers by convection, is evidence of its recent creation there by s-process neutron capture, unconnected with the nuclear fusion in their deep interiors that provides their energy.

Because of the increasing nuclear charge with increasing atomic number, stars need more than about eight times the mass of the Sun to crunch their cores to the higher temperatures and densities required to fuse elements heavier than helium. These more massive stars, although rare at any given moment in the Galaxy, are much more important for increasing the metallicity in the Galaxy because they come and go quickly, producing and dispersing heavier elements on much more rapid evolutionary timescales some two to three orders of magnitude shorter than those for less massive stars.

$$M > 8\,M_\odot$$

When they leave the main sequence, stars having a mass greater than about 8 $M_\odot$ (spectral types O and early B) become *supergiants*, giants on steroids (recall that the surface of the red supergiant Betelgeuse, one of the largest stars known, would extend nearly to Jupiter if it replaced our Sun at the center of the Solar System). After a core of carbon and oxygen is left from hydrogen and helium burning as in lower-mass stars, these more massive stars reach high enough core temperatures ($T \simeq 5 \times 10^8$ K) for the heavy-ion fusion reactions, *carbon burning*:

$$^{12}C + {}^{12}C \rightarrow {}^{20}Ne + {}^4He$$
$$\rightarrow {}^{23}Na + p$$
$$\rightarrow {}^{24}Mg + \gamma,$$

among various other reaction products (where Na and Mg are the chemical symbols for sodium and magnesium, respectively), and *oxygen burning* ($T \simeq 10^9$ K), which includes the following reactions:

$$^{16}O + {}^{16}O \rightarrow {}^{28}Si + {}^4He$$
$$\rightarrow {}^{31}P + p$$
$$\rightarrow {}^{31}S + n$$
$$\rightarrow {}^{32}S + \gamma,$$

(where Si, P, and S denote the elements silicon, phosphorus, and sulfur). Indeed, their higher temperatures and associated energy production rates drive the high luminosities responsible for their greatly extended supergiant sizes. Thus do old ashes become new fuel. Note that higher temperatures are required to fuse the higher proton number nuclei which have correspondingly higher Coulomb barriers. Notice also the rich display of stellar alchemy taking place in these high-mass stars. The words of the English Romantic poet and painter William Blake, "What is now proved was once only imagin'd," would certainly apply to these nuclear transmutations, "proving" the age-old dream of alchemy. And there's much more yet to come.

At these high temperatures, *photodisintegration*—the breakup ("disintegration") of heavy nuclei by high-energy photons ("photo"), the nuclear analogue of the photoionization of atoms that occurs at temperatures a million times lower (reflecting, again, the factor of a million separating nuclear from chemical energies)—such as occurs during the shell "burning" of the neon and silicon produced during carbon and oxygen burning (Fig. 3.17),

$$\gamma + {}^{20}Ne \rightarrow {}^{16}O + {}^4He$$
$$\gamma + {}^{28}Si \rightarrow {}^{24}Mg + {}^4He,$$

releases helium nuclei that undergo successive alpha captures in a series of reactions commencing with silicon,

$$^{28}Si + {}^4He \rightleftarrows {}^{32}S + \gamma$$
$$^{32}S + {}^4He \rightleftarrows {}^{36}Ar + \gamma$$
$$^{36}Ar + {}^4He \rightleftarrows {}^{40}Ca + \gamma$$
$$\vdots$$
$$^{52}Cr + {}^4He \rightleftarrows {}^{56}Ni + \gamma,$$

Fig. 3.17 The onion-like shell structure of the core of a massive star that has evolved through silicon burning. The iron core is surrounded by shells of lighter elements produced through a sequence of hydrostatic burning. At this scale, with the core shown here roughly the size of Earth, the outer layers of the star, primarily inert hydrogen, extend 100,000 times farther, roughly the distance from the Sun to Jupiter. (*Wikimedia Commons*, https://commons.wikimedia.org/wiki/User:Rursus, https://creati vecommons.org/licenses/by-sa/3.0/)

(where Ar, Ca, Cr, and Ni are the symbols for the chemical elements argon, calcium, chromium, and nickel). Note the reverse arrows indicating that the competing photodisintegration and alpha-capture reactions proceed in both directions at these high temperatures, ensuring a state of thermodynamic equilibrium. As Donald Clayton warns, "A stellar center that has passed through the fiery furnace of the [equilibrium] process seems headed for a catastrophic holocaust" [13, p. 542]. Element abundances strongly suggest that truncated silicon burning is an important component of stellar nucleosynthesis in massive stars.

Thus, for most stars in this mass range, silicon burning and neutron capture produce a host of heavy nuclei centered in the iron group near the ^{56}Fe peak of the nuclear binding energy per nucleon curve (recall Fig. 3.10); some stars having a mass at the lower end of this range (8–11 $M_\odot$) halt nuclear burning when their cores become degenerate and end up as oxygen-neon-magnesium (ONeMg) white dwarfs. Regardless of their particular mass, the very high luminosity of stars in this mass range drives very strong metal-enriched stellar winds such as those associated with Wolf-Rayet stars

(Fig. 3.18), stars, named after their two nineteenth-century French discoverers. Laboratory studies of isotopic abundances and mineral inclusions in meteorites provide direct and detailed information on the graphite- and silicate-rich dust particles formed in stellar winds (and, as we'll see below, in supernovae explosions) that drift throughout the Galaxy as interstellar dust grains before being incorporated into new stars—and possibly planets and people.

Because carbon, oxygen, and silicon burning take place at such high temperatures—and hence high reaction rates—and because they produce nuclei having masses progressively nearest the iron peak of the binding energy curve, less and less energy is generated for a given mass of fuel, resulting in less time spent during each succeeding reaction sequence. For example [12, p. 131], for a 25 $M_\odot$ star, the main-sequence hydrogen-burning lifetime is roughly 10 million years, helium burning lasts about a million years, carbon burning only a few hundred years, oxygen burning takes just half a year, while silicon burning lasts just one day![25]

Fig. 3.18 The star Wolf-Rayet 124 (WR 124) and the gas and dust it expelled, captured by the James Webb Space Telescope. The characteristic star-like diffraction spikes are caused by the telescope. (*Photograph courtesy of NASA, ESA, CSA, STScI, Webb ERO Production Team, public domain*)

---

[25] As Gribbin points out [10, p. 127], "In the 1960s and 1970s, when astronomers first developed computer programs to calculate how these changes would occur, it actually took much longer to run the programs than it does for the star to change its structure."

The central iron core continues to grow and approaches the Chandrasekhar mass M_{Ch} as temperatures climb to 10^{10} K, high enough for *nuclear photodisintegration* by energetic photons:

$$\gamma + {}^{56}\text{Fe} \rightarrow 13\,{}^{4}\text{He} + 4n - 124 \text{ MeV},$$

where the minus sign indicates an *endo*thermic (energy-*absorbing*) reaction leading to an "iron catastrophe." The helium nuclei are further photodisintegrated into their constituent protons and neutrons in a reaction that is the reverse of the energy-producing, main-sequence, hydrogen-burning reactions:

$$\gamma + {}^{4}\text{He} \rightarrow 2p + 2n - 28.3 \text{ MeV},$$

another endothermic reaction, absorbing the 28.3 MeV binding energy of the helium nucleus. Altogether, the core absorbs a total energy of about 10^{45} J, equivalent to roughly ten times the energy radiated by our Sun over its entire 10-billion-year main-sequence lifetime (that's A LOT of energy; recall Note 15).[26] Clearly, photodisintegration constitutes a highly endothermic undoing of the entire history of nuclear fusion in massive stars.

And so, we have another "just right" Goldilocks not-too-hot, not-too-cold scenario: nuclei need at least several million K to power the stars and to fuse heavier elements but no more than a few billion K to survive, yet another cosmic coincidence that works out well for life, the Universe, and everything.

Electron-capture neutronization (try that one on your friends!) by high-energy electrons,

$$e + p \rightarrow n + \nu,$$

depletes the core electrons and hence their supporting degeneracy pressure, leading to core collapse on the *free-fall timescale*,[27] which at these high densities and hence strong gravity takes *less than one second* of time, and carries

---

[26] The energy of the core is reduced by $(124 + 13 \times 28.3)/56 \simeq 8.8$ MeV per nucleon, so (recall Note 17) with 10^{57} nucleons per solar mass ($\simeq M_{Ch}$), the core energy is reduced 8.8 MeV/nucleon $\times$ 10^{57} nucleons $\times$ 1.602×10^{13} J/MeV $\simeq 1.4 \times 10^{45}$ J.

[27] For most of their lives, stars maintain a balance—*hydrostatic equilibrium*—between the inward force of self-gravity and the outward force exerted by pressure gradients, typically thermal gas pressure and/or photon radiation pressure for main-sequence stars, degeneracy pressure later in life. Without this support, a star (or stellar core) would collapse as if in free fall on a timescale—the *free-fall timescale* t_{ff} (also known as the *sound-crossing timescale*, the time it takes sound, a pressure fluctuation, to reestablish pressure equilibrium)—that can be estimated from simple physics. The distance R an object falls in a gravitational field $g \simeq GM/R^2$, as established by a spherical mass M having a radius R, in a time t_{ff} is $R = \frac{1}{2}gt_{ff}^2$, a result familiar to students of elementary physics. For a uniform distribution of mass, the mass density ρ = mass/volume = $M/(4\pi R^3/3)$. Combining this with the

away an energy by the neutrinos comparable to that lost by photodisintegration and neutronization combined. At these high densities and temperatures, neutrinos are produced also through the annihilation of electron-positron pairs ($e + e^+ \rightarrow \nu + \bar{\nu}$) and via the photoneutrino process ($\gamma + e \rightarrow e + \nu + \bar{\nu}$). The rate of energy escaping the star in the form of neutrinos can exceed the star's luminous energy by a factor of a million or more.

Thus, most nucleons become neutrons, and, unless the core mass is greater than about two to three times that of the Sun (in which case a black hole is formed), a *neutron star* having a radius of only about 10 km and a density comparable to that of nuclear matter forms as the core collapses and the rest of the star is blown off in one of the most energetic explosive events in the Universe (thereby breaking, in a most violent and spectacular fashion, poet Robert Frost's "calm of heaven"): a neutrino-driven *core-collapse (Type II) supernova* powered by the decrease in the core's gravitational binding energy

$$E_B \simeq GM^2/R = 3 \times 10^{46}(M/M_\odot)^2(10 \text{ km}/R) \ J,$$

more than an order of magnitude greater than the energy lost by photodisintegration and neutronization combined.[28] Degenerate neutrons, rather than

---

equations for g and R gives $t_{ff} \simeq (3/2G\pi\rho)^{1/2} \sim (G\rho)^{-1/2}$, a result that differs very little from the exact solution $t_{ff} = (3\pi/32G\rho)^{1/2}$ and follows directly from dimensional analysis as the only combination of the gravitational constant G and mass density ρ having the dimension of time. For the Sun, $\rho = 1.4$ g/cm^3 and $t_{ff} \simeq 30$ min, so that without internal pressure support, the Sun would collapse to a point in half an hour; clearly, the Sun, like all main-sequence stars, is in hydrostatic equilibrium. For a white dwarf, $\rho \simeq 10^7$ g/cm^3 so $t_{ff} = 1$ s, about the same time it takes the core of a massive star to collapse as it loses pressure support and explodes as a core-collapse supernova, which, although much denser than a white dwarf, generates neutrinos that exert pressure and slows the collapse. At the other end of the stellar life cycle, t_{ff} is about a million years for a protostellar star-forming cloud having a density of 10^4 particles/cm^3.

[28] In 1934, just two years after Chadwick's discovery of the neutron (recall Note 10 of Chap. 2), Fritz Zwicky (1898–1974), an abrasive, Bulgarian-born, Swiss astrophysicist working at the California Institute of Technology along with the German-born American astronomer Walter Baade (1893–1960), were the first to propose that a supernova (their term) is the observational manifestation of the transition of a dying massive star into a neutron star, and that, furthermore, these extremely energetic events are the source of cosmic rays. Named to convey the sense of a very bright nova, both terms are now known to be a misnomer in that these are dying or dead stars, not "new" stars in the sense of the Latin "nova." These so-called core-collapse supernovae are now labelled Type II to distinguish them from Type I accretion-induced supernovae which (as we'll see in the next section) involve the disruptive explosion of a white dwarf star when it accumulates matter from a binary companion, pushing it over the Chandrasekhar mass limit.

An important historical supernova occurred in 1572 when the eccentric Danish nobleman Tycho Brahe observed what he thought was a new star ("nova stella") in the constellation Cassiopeia, but which was actually the phenomenon we now know as a supernova, a massive explosion of a dying massive star. After several nights of carefully observing this stellar intruder and detecting no parallax—no change in its position on the sky relative to neighboring stars as would be expected if it were a nearby object—Tycho concluded that "this new star is… among the fixed stars." (Indeed, modern measurements on the remnant of this star indicate that it was in fact a very distant supernova

electrons, provide the pressure support against gravity: the core has collapsed to a giant nucleus of neutrons. More energetic explosions ("*hypernovae*") may be driven by strong magnetic fields produced by rapid rotation, leaving behind a highly magnetized neutron star ("*magnetar*") or a central black hole ("*collapsar*"). Neutron stars and black holes, along with white dwarfs, are discussed in more detail in the following section sampling the stellar graveyard.

At maximum brightness, core-collapse supernovae rival the brightness of an entire galaxy—100 billion suns—before fading over several months (Fig. 3.19), and eject most of the exploding star's mass into the surrounding interstellar medium at speeds up to several percent of the speed of light thanks to the pressure exerted by the escaping neutrinos which carry away 99% of the total energy associated with the supernova. The kinetic energy of the ejected material, *a lot* of mass moving *very* fast, nevertheless amounts to only 1% of the total energy released, and the luminous energy, produced primarily by the decay of freshly synthesized radioactive elements, despite being as bright as these stellar explosions are, is, at most, a mere 0.1% of the total energy. A shock wave accompanying the "bounce" following core collapse to Pauli-exclusion densities expands outward through the star initiating further element synthesis beyond the iron group (recall Fig. 3.14) and drives an expanding shell of gas and dust—star dust from star guts—observed as a *supernova remnant* (Fig. 3.20). Besides being major producers and distributors of the chemical elements, expanding supernova shock waves trigger the formation of new stars and are the main source from outside the Solar System of cosmic rays—high-energy particles, mainly protons. accelerated to near light speed ("ray" is another misnomer: because of their penetrating power, they were initially thought to be a form of electromagnetic radiation).

In one of the most exciting astrophysical events of the twentieth century, the sudden brightening of a star 170,000 light years away in the Large Magellanic Cloud (LMC), a small irregular satellite galaxy of our Milky Way, was observed on 23 February 1987 and was quickly confirmed as a core-collapse supernova, the brightest supernova detected since the invention of the telescope. Designated SN 1987A (see Fig. 3.21), this "gift from the

---

some 10,000 light years from Earth.) *This was a significant finding, directly challenging the Aristotelian doctrine of an unchanging, immutable Cosmos beyond the sphere of the Moon.* The last supernova directly observed in the Milky Way was Kepler's Supernova in 1604, appearing not long after Tycho's Supernova, both of which were visible to the naked eye. The remnants of more than 300 supernovae have been found in our Galaxy, and observations of supernovae in other galaxies suggest they occur on average about two or three times per century per galaxy, so we're long overdue for one in our neighborhood. The most recent naked-eye supernova was SN 1987A, the explosion of a blue supergiant star in the Large Magellanic Cloud, a satellite galaxy of our Milky Way.

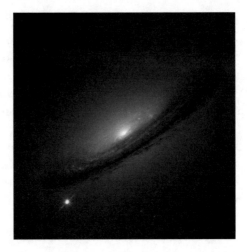

Fig. 3.19 Hubble Space Telescope image of supernova 1994D (SN 1994D), the bright "star" in the lower left, shining as bright as the entire central region of its host galaxy NGC 4526 some 55 million light years from home. (*Photograph courtesy of NASA/ESA, The Hubble Key Project Team and The High-Z Supernova Search Team, public domain*)

heavens" has become the most carefully studied supernova to date, indeed "one of the most extensively observed objects in the history of astronomy... observed in all wavelengths from gamma rays to radio" [14, p. 436]. It stands as "an unequivocal confirmation of explosive nucleosynthesis [that] demonstrated beyond reasonable doubt that heavy elements are indeed cooked and dispersed by supernovae" [15, p. 119]. Its *light curve*—a measure of the change in its brightness over time—is consistent with the decay chain of radioactive isotopes in the ejecta, in particular the β-decay of nickel, produced in the wake of the shock front marching outward through the star, to cobalt (Co, with a half-life of 6.1 days) and eventually to iron (77.7-day half-life; these short half-lives prove these elements were synthesized by the supernova and not preexisting in the star):

$$^{56}\text{Ni} \rightarrow {}^{56}\text{Co} \rightarrow {}^{56}\text{Fe}.$$

Spectroscopic evidence indicates that roughly 0.07 $M_\odot$ of ^{56}Ni, which has the greatest binding energy per nucleon of all nuclei having an equal number of protons and neutrons, was produced in the explosion which ejected a total mass of approximately 15 $M_\odot$. Gamma-ray (γ) emission lines from the cobalt decay ($^{56}\text{Co} \rightarrow {}^{56}\text{Fe} + e + \gamma$) were observed for the first time. Existing observations of the region before the supernova show that the

Fig. 3.20 The Crab Nebula (catalogue designations M1, NGC 1952, Taurus A, SNR 1054), located some 6,500 light years away in the constellation Taurus, is an 11-light-year-wide remnant of a supernova explosion that was recorded by Chinese astronomers as a "guest" star visible in the daytime sky for nearly a month in our year 1054, making it the first astronomical object identified with a historical supernova. Nearly a thousand years later, it's still expanding at a rate of almost 150 km/s. A pulsar—a rapidly spinning neutron star—embedded in the center of the nebula (see Fig. 3.25) has a mass equal to the Chandrasekhar mass and powers its luminous energy output of nearly a million suns (10^5 $L_\odot$). The colors in the image indicate the different elements that were expelled during the explosion: the orange filaments are the tattered remains of the star and consist mostly of hydrogen, the blue in the outer part of the nebula represents neutral oxygen, green is singly-ionized sulfur, and red denotes doubly-ionized oxygen. The eerie blue glow near the center of the nebula is highly polarized synchrotron radiation produced by relativistic particles spiraling in a magnetic field. (*Photograph courtesy of NASA/ESA/ASU/J. Hester and A. Loll, public domain*)

progenitor was a blue supergiant that had a main-sequence mass of about 20 $M_\odot$ and an age of roughly 11 million years.

Certainly, the most exciting observation of SN 1987A was that of the burst of neutrinos from the supernova serendipitously recorded at Japan's Kamiokande II and Ohio's Irvine-Michigan-Brookhaven Cerenkov detectors (after they had passed through Earth!), the first time that neutrinos had been detected from an astronomical source other than the Sun, thus marking the birth of extrasolar neutrino astrophysics, a new window on the Universe. (These detectors were designed to look for proton decay predicted by grand unified theories—GUTs—of physics discussed in Chap. 2.) Together, a total of 20 neutrinos were detected out of the 30 billion that passed through every square centimeter of Earth (recall that neutrinos rarely interact with matter and are therefore difficult to detect), arriving with energies consistent with a

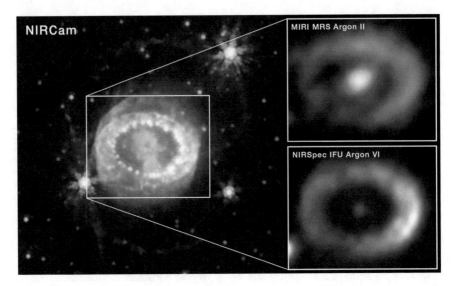

Fig. 3.21 Observations by the recently launched James Webb Space Telescope of infrared emission from the site of SN 1987A provides the best evidence yet for emission from a neutron star. At *left* is a near-infrared image of the remnant of SN 1987A, some 35 years after it exploded. The two images on the *right* show emission from singly (*top*) and multiply (*bottom*) ionized argon, both of which show a strong signal from the center of the remnant indicating a strong source of high-energy radiation, most likely a neutron star. (*Photograph courtesy of NASA, ESA, CSA, STScI, Claes Fransson (Stockholm University), Mikako Matsuura (Cardiff University), M. Barlow (UCL), Patrick Kavanagh (Maynooth University), Josefin Larsson (KTH), public domain*)

blackbody neutrino radiator at a temperature of 5×10^{10} K, a core-collapse temperature some ten million times greater than the surface temperature of the Sun. These observations confirmed the basic theory of core-collapse supernovae (note, again, the agreement between theory and observation, so essential in science) and gave us our first "look" at the formation of a neutron star from a collapsed iron core (recall Fig. 3.21).

The total energy released by the core collapse of SN 1987A exceeded 10^{46} J, indicating that about 0.1 $M_\odot$ of stellar material was converted into pure energy according to Einstein's famous equation, $E = mc^2$ (check this—it's an easy calculation!), with 99% of the energy carried away by neutrinos and less than 0.01% coming out as luminous energy. As supernova specialist Alex Filippenko put it: "Despite being visually spectacular objects … supernovae are, fundamentally, giant generators of neutrinos…. Their light is a sideshow" [15, p. 121].

Looking again at the abundance of the elements (recall Fig. 2.21), we notice several features of interest related to our story of stellar nucleosynthesis. The gradual drop in abundance with increasing atomic number is due to the increasing Coulomb barrier with increasing proton number. As discussed earlier, alpha capture creates nuclei with even numbers of protons and neutrons that have higher binding energies than their "odd" neighbors and is responsible for the abundance peaks for these so-called "alpha elements" such as ^{12}C, ^{16}O, ^{20}Ne, ^{24}Mg, ^{28}Si, ^{32}S, ^{36}Ar, and ^{40}Ca, each one being one alpha capture from the other. Elements heavier than calcium require an excess of neutrons over protons (which are mutually repelled) for stability. The iron-group peak is due to the stability of those nuclei, primarily chromium (Cr), manganese (Mn), Fe, Co, and Ni, having the highest binding energy per nucleon (recall Fig. 3.10); fission rather than fusion becomes energetically favorable beyond this iron-group peak, returning matter to lighter nuclei and providing a rich source of neutrons. Beyond iron and nickel—the majority of the periodic table—the abundance curve has a relatively flat slope with two small rises around barium and lead related to neutron-shell closures for the "magic" neutron numbers $N = 82$ and 126, pointing to the importance of neutron-production processes, the s- and r-processes being the most important—slow and rapid compared to β-decay, the former taking place in the low neutron density environments of red giants, the latter becoming dominant at high neutron densities. It is estimated that about half of the naturally occurring elements in the Universe heavier than iron were created by slow neutron capture and the other half by rapid neutron capture. The required high neutron densities ($\sim 10^{20}$ neutrons/cm^3) for the r-process point to neutron-rich explosive astrophysical processes such as occur in some supernovae but, as realized only recently, mostly in neutron star mergers (discussed in the next section) such as the recently observed merger GW170817, for which the radioactive decay of neutron-rich r-process nuclei was observed along with the spectral fingerprint of freshly synthesized heavy elements.

The relative abundances of various extinct and naturally occurring radioactive isotopes in meteorites and on Earth indicate that the primitive solar nebula was injected, perhaps several times, with r-process-rich elements, and their known rate of decay provides a "clock"—a *nuclear cosmochronology* using nuclear processes to mark time (*chronos*) in the Cosmos—limiting the epoch of the formation of the Solar System to no more than 6 billion years ago. (Radiometric dating indicates the Solar System is about 4.5 billion years old.) As Donald Clayton, an American physicist who literally wrote the book on stellar nucleosynthesis concludes, "natural radioactivities are the most convincing proof of the truth of nucleosynthesis" [13, p. 606, the final paragraph of his book].[29]

And so, as Hans Bethe so succinctly summarized, concluding his 1967 Physics Nobel acceptance speech [16, p. 233]:

> Stars have a life cycle much like animals. They get born, they grow, they go through a definite internal development, and finally they die, to give back the material of which they are made, so that new stars may live.

Because our focus is on understanding and appreciating our cosmic connection through the life cycle of the stars, many of the minutiae of stellar evolution, including the detailed tracings in the HR diagram of the different evolutionary tracks of stars of different mass, many of which pass through various stages of instability, such as those related to outer-layer opacity effects and late-stage shell burning driving the brightness variability of Cepheid, RR Lyrae, and long-period variable stars, have been omitted here. The interested reader can find these and many other related topics omitted from our discussion here examined in several of the suggested readings listed in the backmatter of this book.

3.4   The Stellar Graveyard

> *There is still enough energy in one overlooked star / to power all the heavens madmen have ever proposed.*
>   —final sentence in American author John Updike's "Ode to Entropy"

> *… there is no doubt that neutron star mergers are indeed a major r-process source* [of the chemical elements].
>   —Arcones and Thielemann, "Origin of the Elements" [17, p. 77]

What, you may wonder, is so interesting about dead stars? For starters, the exotic nature of these objects—white dwarfs, neutron stars, and especially black holes—should spark an interest in the stellar graveyard which, not unlike those here on Earth, is filled with the once living. And, as it turns out, many dead stars are, like the character Wesley in the wonderful 1987 film *The Princess Bride*, only "mostly dead," and indeed come alive again in

---

[29] In the preface to the 1984 reprint of his 1968 book, *Principles of Stellar Evolution and Nucleosynthesis*, Clayton happily admits that he "cannot resist author's license to remark that the last paragraph of this book [expressing "optimism for the future of nuclear cosmochronology"] has been abundantly borne out."

certain situations (as the introductory quote suggests).[30] Besides, just as my college anthropology professor pointed out for dead men [18], dead stars do tell tales.

White Dwarfs

The stellar graveyard is populated primarily by white dwarfs, the end stage in life cycle of most stars, those having a main-sequence mass less than about 8 $M_\odot$. With a galactic population estimated at some 10 billion members, white dwarfs are second only to red dwarfs as the most common type of star in the Galaxy.

Perhaps the most famous—and at a distance of only 8.6 light years, certainly the closest—white dwarf is Sirius B, the companion (the "Pup") to the brightest star in the night sky, the "Dog Star" Sirius (from the Greek Σείριος, "the scorcher"), formally labeled α Canis Majoris A, being the brightest star (α) in the constellation of the Big Dog (Canis Majoris).[31] Difficult to see because of a 1000-fold disparity in brightness (Fig. 3.22), it

---

[30] When Wesley, the hero of the story, is thought to be dead, people close to him bring him to Miracle Max, played by Billy Crystal, who examines Wesley's body and announces much to their surprise that he's only mostly dead, which is still a little alive.

It is interesting to recall the prescient words of Donald Clayton [13, p. 591], writing in his book on stellar evolution and nucleosynthesis more than half a century before Arcones and Thielemann reviewed the evidence for r-process nucleosynthesis during neutron star mergers:

> It can be stated with some confidence … that the proper interpretation of the origin of the r nuclei will have considerable astrophysical significance. The unusual demands placed by nuclear physics upon their circumstances of synthesis will certainly put their origin within an unusual and special event. The attempt to satisfy these nuclear demands within a natural astrophysical context is an exciting adventure.

[31] The heliacal rising—the first appearance during the year before the Sun in the early morning sky—of Sirius coincided, at the time of the construction of the Egyptian pyramids, with the annual flooding of the Nile River, which occurred then at about the time of the summer solstice. (Today the waters of the Nile are regulated by the Aswan dam.) Without the life-giving water and nutrient-rich silt, the agriculturally based Egyptian society would have collapsed. So important was this event to the ancient Egyptians that it marked the beginning of the new seasonal year (and continues to be celebrated today as a two-week annual holiday known as Wafaa El-Nil). Unaware that the true cause of the annual flooding was due to a seasonal increase in rainfall at the source of the Nile in equatorial East Africa with the approach of summer, the ancient Egyptians, following the natural human tendency to associate cause and effect to sequentially ordered events, assigned the cause to the star Sirius itself. Thus was born Egyptian astrology. If Egypt is a "gift of the Nile," the Nile is, as depicted in William Blake's 1791 *Fertilization of Egypt*, a "gift of Sirius."

The appearance of the Dog Star—"the scorcher"— with the heat of summer announces, still today, the "dog days" of summer; it was thought that the bright light from this brightest star in the night sky added to the heat of the Sun when the two were near each other in the sky at this time of the year. We have watchdogs to guard our earthly possessions, just as the Dog Star watched for and warned of the annual flooding of the Nile.

was first sighted in 1862 by the American astronomer and famous telescope maker, Alvan Graham Clark (1832–1897), using the largest telescope in the United States at the time, although a companion to Sirius A was suggested by the German astronomer and mathematician Friedrich Wilhelm Bessel (1784–1846) earlier in the century on the basis of its "wobbly" motion across the sky. Gravitationally bound in a binary system, the two stars orbit a common center of gravity about every 50 years. Like most white dwarfs, Sirius B has a mass comparable to that of our Sun (in fact exactly 1 $M_\odot$, half that of its brighter companion), and a size comparable to that of Earth. Its temperature is about 25,000 K compared to 10,000 K for Sirius A.

It's difficult to appreciate the extreme and often bizarre properties of these dead stars. A white dwarf the size of Earth (radius = 6378 km) with the mass of the Sun has an average density of about 2×10^6 g/cm^3, so a level teaspoon (about 5 cm^3) would weigh 10 tons here on Earth (making it very difficult to bake a cake should the recipe call for two teaspoons of white dwarf), and the acceleration due to gravity at its surface (GM/R^2) is more than 300,000 times that of Earth's 9.8 m/s^2 (= 32 ft/s^2), so that level teaspoon on a white dwarf would weigh more than 300,000 times its 10 tons on Earth. Because the gravitational potential energy of a mass m at a height h in a gravitational field g is mgh, the energy required to climb Mount Everest ($h \simeq 10$ km) will

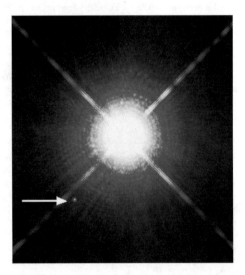

Fig. 3.22 A Hubble Space Telescope image of the bright star Sirius A and its much fainter white dwarf companion Sirius B (lucky for us, indicated by the Creator with an accompanying arrow!). The diffraction spikes and concentric rings are instrumental effects. (*Photograph courtesy of NASA, ESA, H. Bond (STScI), and M. Barstow (University of Leicester), public domain*)

lift you only an inch ($\simeq$ 3 cm) at the surface of a white dwarf. Strong gravity, indeed.

Significant mass transfer from a binary companion star onto a white dwarf close to the critical Chandrasekhar mass (or, more rarely, the gravitational inspiral of binary white dwarfs) can push the star over the Chandrasekhar limit, causing a thermonuclear runaway because, under degenerate conditions, a white dwarf is insensitive to temperature and thus lacks a controlling thermostat, leading to a complete, explosive destruction of the white dwarf— a star-size thermonuclear bomb—an event known as an accretion-induced *Type Ia supernova*, rivaling core-collapse Type II supernovae in energy production. About a quarter of all supernovae are Type Ia; the supernova shown in Fig. 3.19 is a Type Ia event. Type Ia supernovae contribute more than half of the iron-group elements in the Universe (recall Fig. 3.14), leaving the remaining fraction to core-collapse Type IIs; some heavier alpha elements (Si, S, Ar, and Ca) are also produced by Type Ia supernovae, but less than the contribution from Type IIs. Because most Type Ia supernovae involve progenitors all at the Chandrasekhar mass limit, they have a narrow range of maximum intrinsic brightness, and thus serve as useful distance indicators, "standard candles" to probe the far away and hence the long ago (which led to the discovery in the late twentieth century of the accelerated expansion of the Universe and Nobel Prizes in 2011 for the three team leaders).

These accretion-induced supernovae are a much scaled up version of *novae*, cataclysmic variable stars that suddenly and temporarily brighten by up to a factor of 10^6, far less than that of supernovae (whence the prefix "super"). Another celestial misnomer, the name suggests the appearance of a "new" (Latin, *nova*) star, when in actuality, these are dead stars rendered visible by their sudden brightening as if they were "new" stars; because they brighten and fade, the ancient Chinese called them "guest stars," visiting us only briefly in the nighttime sky. Some novae are recurrent on timescales up to hundreds of years, and all are members of binary star systems, which explains their flare ups as the result of the transfer of mass, mostly fresh hydrogen fusion fuel, onto the surface of a hot white dwarf, resulting in a small thermonuclear explosion as the hydrogen ignites, ejecting about 10^{-5} $M_\odot$, stopping the mass transfer temporarily. Mass transfer onto a neutron star or black hole produces bursts of X-rays in a system known as an *X-ray binary*, higher-energy radiation reflecting the stronger gravitational energy of these compact objects. And so it is that these "dead" stars live again after being only "mostly dead." The nucleosynthesis contribution of either of these types of novae is nearly negligible.

Because white dwarfs are dead stars no longer able to generate energy, they slowly cool as their internal thermal energy is slowly radiated away. Because they are supported by degenerate electron pressure, which is independent of temperature, their radius remains constant as they cool. Accordingly, the Stefan-Boltzmann equation introduced in the previous section, $L = 4\pi R^2 \sigma T^4$, predicts $L \sim T^4$, in agreement with observations which place white dwarfs along a narrow diagonal strip in the lower-left corner of the HR diagram (recall Fig. 3.8). White dwarfs evolve down this strip as they cool and become fainter. And, just as every good detective knows that the time of death can often be deduced from the temperature of the corpse, the age of the oldest ("deadest") white dwarf—and hence the minimum amount of time since stars began to form in the disk of our Galaxy—can be estimated from the white dwarf cooling timescale, the time taken to lose its reservoir of thermal energy, quite a long time due to their small size (recall from the Stefan-Boltzmann equation that $L \sim R^2$) and slow heat transfer from their interior. A dramatic drop in the observed luminosity of white dwarfs implies that they began forming about 9 ± 2 billion years ago, nearly as long ago as our Galaxy's globular clusters formed some 12 billion years ago, nearly two billion years after the Universe itself formed. This age sequence—old white dwarfs, older globular clusters, and the oldest of all, the Universe, is beautifully self-consistent, just what we would expect.

But there's more in the stellar graveyard than meets the eye, and we turn now to these even more exotic stellar corpses, many of which are important creators of the chemical elements.

Neutron Stars

The theory of neutron stars—the neutron-rich stellar remnants of supernovae having a mass comparable to the Sun packed into an object the size of a small city, giving them a density similar to that of the atomic nucleus—was developed in 1939 by the American theoretical physicist J. Robert Oppenheimer (1904–1967), who later headed the atomic bomb project during World War II, working with the Russian-Canadian physicist George Volkoff (1914–2000). Oppenheimer and Volkoff showed that the degenerate neutrons supporting a neutron star against gravitational collapse cannot support more than about 2 to 3 solar masses, the so-called *Oppenheimer-Volkoff limit*—a kind of neutron-star Chandrasekhar limit—beyond which there is no known mechanism that can prevent complete gravitational collapse.

Even more so than white dwarfs, it's not easy to appreciate the extreme conditions of neutron stars; indeed, we don't fully understand matter at these high, nuclear densities (a few $\times$ 10^{14} g/cm^3 = a few 100 million tons/cm^3), equivalent to packing all the people on Earth today—and then some—into a teaspoon (so much of humanity being so much empty space). Neutron stars have a radius of about 10 km, much smaller than white dwarfs because the size of a star supported by degeneracy pressure, whether electron or neutron, is inversely proportional to the mass of the degenerate particle [6, p. 84], and the neutron is nearly 2000 times more massive than the electron. The dependence of radius on mass is weak and, in any case, the likely mass range for neutron stars is fairly small, ranging from near the Chandrasekhar mass up to the Oppenheimer-Volkoff limit: most neutron stars with known mass have a mass very close to the Chandrasekhar mass of 1.4 $M_\odot$.

The pull of gravity at the surface of a solar-mass neutron star having a radius of 10 km is over 100 billion times stronger than that on Earth, so that every pound you have here on Earth would weigh 100 billion pounds on the surface of a neutron star (clearly not a place to visit if you're trying to lose weight). And the energy required to climb Mount Everest on Earth that would get you over a one-inch rise on a white dwarf will elevate you on a neutron star only one micron (one-millionth of a meter; about one-hundredth the width of a human hair). (No wonder life forms on the fictional neutron star, Dragon's Egg, are the size of a sesame seed and are restricted by strong surface gravity to movement in only two dimensions along the surface [19].) An object dropped from a height of one meter would strike a neutron star's surface at a speed of half a percent of the speed of light (nearly 4 million mph); if that dropped object were a 7-g marshmallow, it would have an energy equivalent to about four tons of TNT. Clearly extreme conditions. The large ratio of neutron star gravitational energy (~ GM^2/R) to rest energy (Mc^2), $GM/Rc^2 \simeq 0.2$, indicates strong gravity and therefore the need to consider general relativistic effects such as that of strong gravity bending light and slowing down time (for comparison, that ratio is about 10^{-9} for Earth and 10^{-6} for the Sun, and still only 10^{-4} for white dwarfs). Gravity on the surface of a neutron star bends light so one can see 20° to 30° over the horizon, and makes clocks tick two times slower than here on Earth. The most important relativistic effect of gravity is that its attractive nature is strengthened at very high densities and pressures.[32]

---

[32] Einstein announced his *general theory of relativity*, which interprets gravity as a warping of space-time by mass-energy, in 1915, ten years after introducing his *special theory of relativity* addressing the effects of high-speed motion on the properties of space (contraction) and time (dilation) as well as the equivalence of mass and energy ($E = mc^2$). The *gravitational deflection of light* (acting like

The theory of neutron stars received observational support when, while studying the scintillation of point radio sources in the sky in 1967, Jocelyn Bell (now Bell Burnell; b. 1943; Fig. 3.23), a Cambridge University graduate student, discovered the first *pulsar* (*pulsa*ting *r*adio stars, named to be similar to *quasar*, a then-recently discovered *quas*i-stell*ar* radio source, now known to be an extremely luminous active galactic nucleus powered by a supermassive black hole). Bell and her Ph.D. supervisor, Anthony Hewish (1924–2021), initially dubbed them LGM for "Little Green Men," jokingly implying they might be navigational beacons placed throughout the Galaxy by an advanced civilization (which, as you can imagine, caused quite a stir). Hewish shared the 1974 physics Nobel Prize for Bell's discovery; Bell, female and "only" a student, received nothing despite being the actual discoverer. In 1968, just one year after their discovery, the Austrian-American astronomer Thomas Gold, one of the authors of the steady-state model of the Universe (recall Chap. 2), suggested that pulsars are rapidly rotating neutron stars sweeping out, like a lighthouse, a narrow beam of energy focused by their

---

a lens, the presence of mass bends space and hence the path of light) and a *gravitational redshift* and *time dilation* (gravity slows time, including the oscillation period of light, effectively stretching it to longer wavelengths) are consequences of the gravitational warping of space-time. Although Earth's gravity is weak ($GM/Rc^2 \simeq 10^{-9}$; a beam of light passing by Earth will bend only about half an inch in traveling 3000 miles, and a clock upstairs in your house will gain only one second every hundred million years compared to a clock downstairs), extremely sensitive atomic clocks at different altitudes have verified Einstein's prediction. The satellites used in the Global Positioning System (GPS), orbiting higher in Earth's gravitational field which is weaker than at the surface, with the result that their clocks run slightly faster overall than clocks on the ground (about 40 microseconds per day), must incorporate relativistic time dilation corrections to accurately pinpoint positions on Earth, which are calculated by comparing radio-frequency time signals from the satellites with signals from ground transmitters. Einstein would have been impressed. In 1959, Harvard physicists Robert Pound and Glen Rebka detected the gravitational redshift due to Earth's (weak) surface gravity; a measurement made in 2020 detected a 2×10^{-6} ($\simeq GM/Rc^2$) fractional gravitational redshift on the Sun, and a measurement of the gravitational redshift of light from the white dwarf Sirius B, first made in 1971 and refined more recently by the Hubble Space Telescope, found a fractional shift of 3 $\times 10^{-4}$. *Gravitational lensing*, first detected in 1979, appropriately the centennial of Einstein's birth, has become an important tool for detailed studies of the most distant reaches of the Universe and for searching our Galaxy for dark objects such as black holes and extrasolar planets. In 1916 Einstein showed that his field equations have wave-like solutions and predicted that accelerated mass radiates *gravitational waves*, ripples in the fabric of space-time, just as accelerated electric charge radiates electromagnetic waves. Nearly a century after their predicted existence, gravitational waves emitted from the inspiral and eventual merger of two stellar-mass black holes were first directly observed in 2016 by the *L*aser *I*nterferometer *G*ravitational-Wave *O*bservatory (LIGO) in the United States, earning three key LIGO researchers the 2017 physics Nobel Prize. More detections followed, including one the following year from two merging neutron stars.

The confirmation of the deflection of starlight by the Sun during a total solar eclipse in 1919 brought Einstein, carrying "a message of a new order in the Universe" [20, p. 311; recommended reading for more on Einstein and his physics], into the public light and marked the beginning of the perception by the general public of Einstein as a world figure—indeed the beginning of the Einstein legend. No scientist before or since has so completely transcended the role of expert to become a universal icon of reason and a trademark of human intellectual achievement.

strong magnetic fields that are enhanced by core compression during collapse to strengths that can exceed 100 million tesla, a trillion times stronger than Earth's magnetic field (Fig. 3.24).

There are about 3000 known pulsars in the Galaxy and certainly many more that remain undetected due to their narrow beams missing our planet. Although, as we have noted, most stars are members of multiple-star systems, the mass loss accompanying the supernova explosion that leaves behind a neutron star often disrupts these systems, ejecting the pulsar at a high velocity, so that only about 10% remain in binary systems. They spin—and hence "pulse"—with short and remarkably regular periods ranging from milliseconds to seconds, spinning rapidly due to conservation of angular momentum during core collapse (the same principle that enables an ice skater to spin rapidly when "collapsing" her arms close to her chest). The Crab pulsar (Fig. 3.25), discovered in 1968, was the first to be connected with a supernova remnant. It's the youngest known pulsar and therefore not coincidentally the fastest rotating of any normal pulsar, spinning at a rate of about 30 times per second. Pulsars are very accurate clocks, good to 1 part in 10^{16} (the period of a typical pulsar has increased by about 1 s since the age of the dinosaurs), slowly slowing down with age (except for the millisecond pulsars which are sped up by mass transfer from a binary companion) as they slowly convert

Fig. 3.23 Jocelyn Bell Burnell, featured in the *Great Irish Scientists* series in Ireland's Dublin airport. The Crab Nebula, the remnant of a supernova that occurred in the constellation Taurus the Bull in the year 1054 (recall Fig. 3.20), is pictured beside her with superimposed (green) spikes of radio emission that she first detected in 1968 coming from an embedded pulsar that spins about 30 times per second. (*Photograph by the author*)

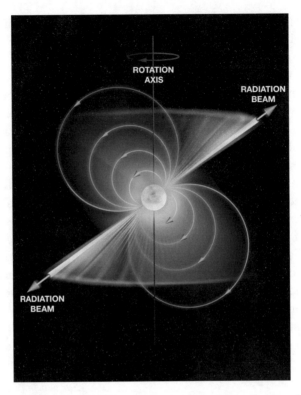

Fig. 3.24 The lighthouse model of a pulsar showing the (*yellow*) radiation beamed along the direction of the dipole magnetic field axis (magnetic field lines in *blue*) which is inclined to the neutron star's (*red*) rotation axis. If Earth lies in the direction of the sweeping beam, a periodic pulse of radiation will be recorded each time the beam sweeps past, much like the periodic flickering of the sweeping beam of light from a lighthouse. (*Photograph courtesy of the National Radio Astronomy Observatory, public domain*)

rotational energy into luminous energy, mostly in the form of highly polarized *synchrotron radiation*—radiation emitted by relativistic electrons as they spiral along magnetic field lines (so named because it was first detected in synchrotron particle accelerators here on Earth). The detailed physics underlying the pulsar emission mechanism is extremely complicated, involving, among other things, the interaction of strong, rapidly-rotating magnetic fields with the surrounding plasma.

The first and best-known of a handful of *binary pulsars*, two neutron stars orbiting each other—another "gift from the heavens"—was discovered by graduate student Russell Hulse (b. 1950) and his thesis advisor Joseph Taylor

Fig. 3.25 A composite image of the Crab Nebula showing the X-ray (blue) and optical (red) images superimposed. Like a lighthouse, the neutron star ejects twin beams of radiation that appear to pulse 30 times per second due to the star's rotation. (*Photograph courtesy of NASA/HST/CXC/ASU/J. Hester et al., public domain*)

(b. 1941) in 1974.[33] Designated PSR1913 + 16 (from its coordinates on the sky), the two objects orbit each other in slightly less than eight hours and are so close together that their orbit would almost fit inside the Sun. The orbit is decaying by about 3.5 m/yr causing a slight but measurable change in the system's orbital period amounting to 75 millionths of a second per year as the system loses energy through the radiation of gravitational waves exactly as predicted by Einstein's theory of general relativity. Until the recent direct observation of gravitational waves (discussed below), this was the best test of general relativity in strong gravitational fields, so important that the pair were awarded the 1993 Nobel Prize in Physics, the first Nobel motivated in part by reference to gravitational physics. Even for the strongest sources—binary black holes or neutron stars, and supermassive black holes at the centers of galaxies—the intensity of gravitational waves is far below that for electromagnetic waves; the Sun radiates as much energy in gravitational waves as a small light bulb does in electromagnetic waves.

---

[33] I was the lucky graduate student assigned to operate the projection system during Hulse and Taylor's presentation of their discovery results to an overflowing audience at the winter 1974 meeting of the American Astronomical Society in Gainesville, FL (I was a Ph.D. student at the University of Florida). Seven years later, in a twist on "Who Got Einstein's Office," as a National Science Foundation summer research associate at the University of Massachusetts, where Taylor and Hulse were working when they made their discovery, I was assigned to Taylor's office which he had just vacated to take a new position at Princeton University.

Nearly a century after their predicted existence, gravitational waves emitted from the inspiral and eventual merger of two stellar-mass black holes were first directly observed in 2015 by the extraordinarily sensitive *L*aser *I*nterferometer *G*ravitational-Wave *O*bservatory (LIGO) in the United States, earning three key LIGO researchers heading a team of more than a thousand physicists and engineers the 2017 physics Nobel Prize. The extreme sensitivity of this detection is related by astronomer Alex Filippenko [15, p. 142];

> The length of LIGO's 4-km [interferometer] arms changed by only about 1/1,000 of the diameter of a proton (which measures only about 10^{-15} meters across), equivalent to the nearest star other than the Sun (4.2 lightyears away) changing its distance by the width of a human hair!

More detections followed, including, two years later, the first from two merging neutron stars (designated GW170817 for the date the "GW" signal was detected), these in a binary system in a distant galaxy about 130 million light years away: neutron stars, like Wesley in *The Princess Bride* and white dwarfs in mass-transfer binaries, are only "mostly dead," and, as we've recently learned, do come back from the dead when they merge with each other releasing tremendous amounts energy and creating, through *r*-process neutron capture in the associated neutron-rich environment, most of the heavy elements in the Universe including the "jewelry store" elements gold, silver, and platinum. "The rapid neutron-capture process needed to build up many of the elements heavier than iron seems to take place *primarily in newton-star mergers*, not [as previously believed] *supernova explosions*" [21, p. 30, emphasis added; see also 22 and, for an updated confirmation, 23].

The optical/infrared transient event powered by the radioactive decay of r-process nuclei heavier than iron, synthesized in the merger of two neutron stars or a neutron star and a black hole that emerges in the ejecta days after the merger, is called a *kilonova* (Fig. 3.26). GW170817 was the first detection of an electromagnetic counterpart to a gravitational wave event, designated AT 2017gfo (AT for *A*stronomical *T*ransient, followed by the year and an arbitrary sequence of letters assigned by the Transient Name Server to uniquely identify the event). This event also marked the first (spectroscopic) identification of an r-process element produced during a neutron star merger: strontium (atomic number 38). Since then, several more r-process elements have been identified in the ejecta including yttrium (Y), tellurium (Te), lanthanum (La), and cerium (Ce; atomic numbers 39, 52, 57, and

Fig. 3.26 An artist's impression showing two neutron stars just as they merge and explode as a kilonova, the main source of very heavy chemical elements in the Universe (recall Fig. 3.14). These events produce both gravitational waves and a short gamma-ray burst, both of which were observed during the 2017 neutron star merger. (*Photograph courtesy of University of Warwick/Mark Garlick,* https://creativec ommons.org/licenses/by/4.0/)

58, respectively); Te was also observed by the recently launched James Webb Space Telescope (JWST) in a kilonova a billion light years away. A total of 16,000 times the mass of Earth in heavy elements is believed to have formed during the GW170817 event, including approximately 10 Earth masses of just the two elements gold and platinum. (Think about how much gold and platinum this is. Incomprehensible.) All of this provides the most direct proof that neutron stars are indeed made of neutron-rich matter.

A short (approximately 2-s duration) *gamma-ray burst* (GRB), designated GRB 170817A, was detected beginning 1.7 s after the gravitational wave merger signal. GRBs were discovered in the 1960s with the US military's Vela spy satellites used to verify nuclear treaties by detecting gamma rays produced during the testing of nuclear bombs. Each day, somewhere in the sky there is a brief burst of gamma rays. Long-duration GRBs, lasting up to a few hundred seconds, are associated with hypernovae, driven by strong magnetic fields produced by rapid rotation during a core-collapse supernova, leaving behind a highly magnetized neutron star (magnetar)[34] or a central black hole (collapsar). Short-duration GRBs, lasting two seconds or less, are produced during the merger of inspiraling neutron stars (as occurred for GW170817/ AT 2017gfo, catalogued as GRB 170817A. Releasing about 10^{44} J of energy, as much as the Sun will release over its entire main-sequence lifetime, they are as bright in gamma rays as the planet Venus is in visible light, despite being at cosmological distances.[35] Observed across the electromagnetic spectrum, these objects mark a significant breakthrough for multi-messenger astronomy.[36]

The discovery of the radioactively powered kilonova AT 2017gfo associated with the short-duration gamma-ray burst GRB 170817A and the gravitational wave source GW170817 provided the first direct evidence supporting binary neutron star mergers as crucial astrophysical sites for the synthesis of heavy elements through r-process nucleosynthesis, demonstrating that these events are the dominant contributor to the production of heavy r-process elements in the Universe. The mergers of binary compact objects such as neutron stars and black holes remain of central interest as sources high-frequency gravitational waves, GRBs, and heavy element nucleosynthesis via rapid neutron capture across a broad range of atomic mass across the Universe.

---

[34] Recent observations and modeling of supernovae light curves suggest that some supernovae may have magnetar central engines powered by the decay of enormous internal magnetic fields providing energy for several minutes to months after the explosion. Rarely, magnetars flare up releasing up to 10^{40} J of energy in gamma rays in less than a second, for a gamma-ray luminosity (10^{40} W) equivalent to 25 trillion times the solar luminosity. Magnetar magnetic fields can be several thousand times stronger than those of typical neutron stars, already strong enough, and thus highly unstable to reconnection annihilation and hence flaring.

[35] This is the equivalent of 10^{51} ergs of energy as measured in the cgs (centimeter-gram-second) system of units, and defines a new "astronomical" energy unit, the *foe*—for ten raised to the *f*ifty-one *ergs*—equal also to the *bethe* (B), named in honor of Hans Bethe, convenient units for expressing the tremendous amount of energy released by a supernova.

[36] The discovery and subsequent observations of GW170817 were given the "Breakthrough of the Year" award for 2017 by the weekly journal *Science*. Scientific interest in the event was enormous, with dozens of preliminary papers published the day of the announcement. The interest and effort were global: the paper describing the multi-messenger observations [24] is coauthored by almost 4000 astronomers—about one-third of the worldwide astronomical community—from more than 900 institutions, using more than 70 observatories in space and on all seven continents, and has been downloaded more than 650,000 times as of April 2024.

Black Holes

In 1916, the year he died of a horrible autoimmune disease contracted while serving on the Russian front during World War I, the German astronomer Karl Schwarzschild (1873–1916) became the first person to solve Einstein's general relativistic field equations for the special case of an isolated point mass M. He found that the space–time curvature becomes infinite— a net representing space–time sags so far that it "reconnects" and pinches off from our Universe the sagging section of the net (Fig. 3.27)—at a radial distance

$$R_S = 2GM/c^2 \approx 3 \text{ km} (M/M_\odot),$$

now called the *Schwarzschild radius* in his honor. As the French polymath Pierre-Simon de Laplace showed at the end of the eighteenth century, this same equation can be derived (coincidentally) from the Newtonian equation for the escape velocity from the surface of a mass M having a radius R_S,

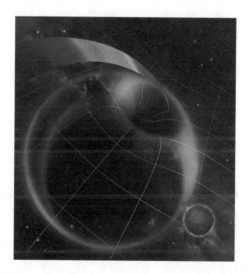

Fig. 3.27 In this 1971 painting titled *Artist's Rendition of a Black Hole* by the German-born American artist Helmut Karl Wimmer, the warped space surrounding a black hole draws in matter and a rainbow of light from a companion star. As the American theoretical physicist John Archibold Wheeler, the originator of the term "black hole" and the person largely responsible for reviving interest in general relativity in the United States after World War II, once said, summarizing Einstein's geometrization of gravity in his theory of general relativity, "Matter tells space how to curve, and curved space tells matter how to move." Eddington likened the relativistic distortion of space to the "world reflected in a polished door-knob." (*Credit Wikimedia Commons by Helmut Karl Wimmer, CC BY-SA 4.0*)

by setting the escape velocity equal to c, the speed of light. Since nothing can travel faster than light—"impossible ... and certainly not desirable, as one's hat keeps blowing off," the American actor and director Woody Allen affirmed [25]—and light itself cannot escape from the region within R_S, the Schwarzschild radius is also referred to (in this case) as the "event horizon" because no information on anything—no "event"—within this region can be communicated to the outside.[37] Such a region, gravity's "fatal attraction," would appear as a *black hole*, a term coined in 1967 by the American theoretical physicist John Archibald Wheeler (1911–2008; whose students include Richard Feynman), an idea initially rejected by Einstein—and by Eddington who thought "that there should be a law of Nature to prevent the star from behaving in this absurd way" [26, p. 38]. With black holes, gravity, the driving force of stellar evolution, has progressed to its ultimate end: infinite compression.

The large ratio of the gravitational binding energy to rest-mass energy of a black hole at the Schwarzschild radius, $GM/R_S c^2 = \frac{1}{2}$, indicates strong gravity and therefore the need to consider general relativistic effects; recall that for neutron stars and white dwarfs, this ratio was about 0.2 and 10^{-4}, respectively. Strong tidal forces—differential gravity—associated with black holes will kill you if you get too close: falling into a black hole feet first, the pull on your feet will be so much stronger than the pull on your head that you'll be stretched out into an uncomfortably thin configuration (Fig. 3.28; the technical term for such a diabolical demise is "*spaghettification*": to be made long and thin like spaghetti).

Research laying the foundations for a general relativistic theory of stellar structure and evolution by Oppenheimer and others in the 1930s suggested that neutron-degenerate objects more than about two or three times the mass of the Sun undergo complete gravitational collapse to a black hole. Such objects are found in the cores of massive stars, those stars having a main-sequence mass greater than about 20–25 $M_\odot$; neutron stars are left behind from stars having a main-sequence mass of roughly 10–20 $M_\odot$; white dwarfs, as we have seen, are the remnants of stars less massive than about 8 $M_\odot$. From the equation for the Schwarzschild radius, one can see immediately that stellar-mass black holes are very compact: $R_S = 30$ km for a star having

---

[37] Interestingly, Schwarzschild's name in German means "black shield," which is essentially what the event horizon is, "shielding" the space outside from the black hole within. One of his sons, Martin Schwarzschild (1912–1997), became a well-known American astronomer. For a mass M, dimensional analysis shows that the only combination of the two physical constants relevant to general relativity, the Newtonian gravitational constant, G, and the speed of light, c (only G is relevant to classical Newtonian gravitational theory), having the dimension of length is GM/c^2, exactly half the Schwarzschild radius.

Fig. 3.28 The author photographing himself in front of a mirror at Chicago's Adler Planetarium, carefully curved to mimic one's appearance falling feet first into a black hole. I'm tall but not this tall here on Earth. (*Photograph by the author, far from black holes*)

a mass 10 times that of the Sun; Earth would have to be compressed into a sphere less than an inch across to become a black hole. Astronomers have found several candidate stellar-mass black holes, the first being Cyg X-1 (the brightest X-ray source in the constellation of Cygnus the Swan) having a mass, determined by the motion of its spectroscopic binary companion, of at least 8 $M_\odot$, as well as supermassive (M > several million to several billion $M_\odot$) black holes lurking at the center of most galaxies including our own. Relativistic astrophysics is now a recognized and important field of physics and astronomy.

Because most stars are members of binary or high-order multiple systems, black holes are sometimes paired with an evolved, mass-transferring companion star (as is Cyg X-1). Due to the orbital motion of the system, the accreting material goes in orbit around the black hole forming an *accretion disk* which heats up to about 10 million K as the kinetic energy of the accreting material is thermalized and radiated away (Fig. 3.29). At this temperature, the peak of the radiation (as per the Wien displacement law) is in the X-ray portion of the electromagnetic spectrum (whence "X-ray binary"). This radiation, originating outside the event horizon, can be detected revealing the monster in the middle. A lower limit to the mass of a black hole (or, for that matter, any unseen object such as a neutron star or a planet—exoplanets are detected this way) can be determined from the motion of a companion star; if this lower limit exceeds the Oppenheimer-Volkoff limit for a neutron star, the unseen companion must be a black hole. Thus does gravity reveal the monster.

And in more ways than one: the first detection of gravitational waves was made on 14 September 2015 by the Laser Interferometer Gravitational-Wave Observatory (LIGO) in Washington State and Louisianna. Two black holes 1.3 billion light years away, having masses of 30.6 and 35.6 suns, merged into a single black hole of about 63 solar masses, with the difference of 3 solar masses being converted into the energy radiated by the gravitational waves in accordance with Einstein's famous equation, $E = mc^2$.

It is fitting to end our discussion of the end stages of the life cycle of the stars with the words ending A. C. Phillips' concise and splendid treatment of *The Physics of Stars* [12, p. 195]:

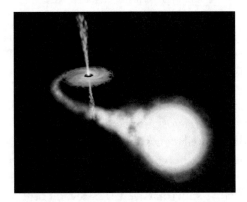

Fig. 3.29 Artist's rendition of a black hole with an orbiting binary companion star transferring mass towards the black hole, forming an accretion disk. Blowtorch-like jets (shown in *blue*) are streaming away from the black hole at near the speed of light, in a scaled-down version of much more massive black holes found in the cores of extremely active galaxies called quasars, giving these systems the name "*microquasars.*" (*Photograph courtesy of ESA, NASA, and Felix Mirabel (French Atomic Energy Comission and Institute for Astronomy and Space Physics/Conicet of Argentina), public domain*)

Gravity is the driving force for stellar evolution. It leads to the formation of a star and to temperatures which make thermonuclear fusion possible. The energy released by fusion only serves to delay the gravitational contraction of the matter inside the star. The end-point may be a white dwarf or a neutron star, stars in which cold matter resists the force of gravity. An alternative end-point is a black hole in which gravity is completely triumphant. This outcome is neat and tidy: nothing is left of the collapsed matter apart from an intense gravitational field.

* * *

Nearly a century ago, in the final sentence of his 1926 groundbreaking book on *The Internal Constitution of the Stars*, the English astrophysicist Sir Arthur Eddington, the first to apply the basic laws of physics to investigate the interior of stars, giving us our first true understanding of stellar processes, appreciated that, unlike the complexities of matter at terrestrial temperatures, "which are likely to prove most difficult to unravel," the "high temperature in the interior of a star is not an obstacle to investigation but rather tends to smooth away difficulties" (recall Richard Feynman's feelings on this mentioned at the beginning of the chapter), and so, Eddington continued, "it is reasonable to hope that in the not too distant future we shall be competent to understand so simple a thing as a star" [27, p. 393]. In his time, one of the "clouds obscuring the theory of the structure and mechanism of the stars" was "the failure of our efforts" to understand the source of stellar energy, what it is that makes stars shine. A little more than a decade later, that outstanding problem—and so much more—had been solved. We now appreciate stars, distant points of twinkling light punctuating the blackness of the night sky, as giant thermonuclear fusion reactors, nuclear pressure cookers cooking the ingredients for life, the Universe, and everything.

The discovery of the life cycle of a star and, along the way, the source of stellar energy and the accompanying synthesis of the elements in stars, has been called [28, p. 610] "the most profound discovery of the whole scientific endeavour." The four most abundant molecule-forming elements in the Universe—hydrogen, carbon, oxygen, and nitrogen—the latter three produced during stellar nucleosynthesis, are the same four most common elements involved in the chemistry of life. Gribbin continues [28, pp. 611–612]:

> We are made out of exactly the raw materials which are most easily available in the Universe. The implication is that Earth is not a special place, and that life forms based on [these elements] are likely to be found across the Universe, not just in our Galaxy but in others. It is the ultimate removal of humankind

from any special place in the cosmos, the completion of the process that began [in 1543] with Copernicus and *De revolutionibus*.

We turn now to the next chapter for further discussion of this cosmic connection—of dust to us—and its implications for life in the Universe.

References

1 R. P. Feynman, R. B. Leighton, M. Sands, *The Feynman Lectures on Physics,* vol. 1, "Mainly Mechanics, Radiation, and Heat"; Chapter 3, "The relation of Physics to other sciences"; section 3–4, "Astronomy" (Addison-Wesley Publishing Company, Reading, MA, 1963)

2 S. W. Huggins, L. Huggins (eds.), *Scientific Papers of Sir William Huggins, footnote added to 1864* "On the Spectra of Some of the Fixed Stars" (William Wesley and Son, London, 1909)

3 J. Al-Khalili, *The World According to Physics* (Princeton University Press, Princeton and Oxford, 2022)

4 Wikipedia, "Pleiades in Folklore and Literature" https://en.wikipedia.org/wiki/Pleiades_in_folklore_and_literature Accessed 5 May 2024

5 S. Weinberg, *The First Three Minutes: A Modern View of the Origin of the Universe* (Basic Books, New York, 1977; rev. ed. 1993)

6 D. Maoz, *Astrophysics in a Nutshell* (Princeton University Press, Princeton and Oxford, 2007)

7 A. S. Eddington, *Stars and Atoms* (Oxford University Press, Oxford, 1927)

8 H. Kragh, *Quantum Generations: A History of Physics in the Twentieth Century* (Princeton University Press, Princeton, 1999)

9 H. A. Bethe, "Energy production in stars," Phys. Rev. **55**, 434–456 (1939). https://journals.aps.org/pr/pdf/10.1103/PhysRev.55.434

10 J. Gribbin, *Stardust* (The Penguin Group, Chatham, UK, 2000)

11 E. M. Burbidge, G. R. Burbidge, W. A. Fowler, F. Hoyle, "Synthesis of the Elements in Stars," Rev. Mod. Phys. **29**, 547–650 (1957). https://doi.org/10.1103/RevModPhys.29.547

12 A. C. Phillips, *The Physics of Stars*, 2nd edn. (Wiley, Chichester, UK, 1999; orig. publ. 1994)

13 D. D. Clayton, *Principles of Stellar Evolution and Nucleosynthesis* (University of Chicago Press, Chicago & London, 1983; orig. publ. McGraw-Hill, New York, 1968)

14 D. Arnett, *Supernovae and Nucleosynthesis: An Investigation of the History of Matter, from the Big Bang to the Present* (Princeton University Press, Princeton, 1996)

15 A. V. Filippenko, "Stellar Explosions, Neutron Stars, and Black Holes," in *Origin and Evolution of the Universe: from Big Bang to Exobiology*, 2nd edn.,

ed. by M. A. Malkan, B. Zuckerman (World Scientific, Singapore, 2020), pp. 99–147

16 H. Bethe, *Nobel Lectures: Physics 1963–1970* (Elsevier Publishing Company, Amsterdam, 1972)

17 A. Arcones, F.-K. Thielemann, "Origin of the Elements," Astron. Astrophys. Rev. **31**, 1 (2023). https://doi.org/10.1007/s00159-022-00146-x

18 W. A. Maples, M. Browning, *Dead Men Do Tell Tales: The Strange and Fascinating Cases of a Forensic Anthropologist* (Doubleday, New York, 1994)

19 R. Forward, *Dragon's Egg* (Del Ray, New York, 1980)

20 A. Pais, *Subtle is the Lord ... The Science and the Life of Albert Einstein* (Oxford University Press, Oxford, 2005; orig. publ. 1982)

21 A. Frebel, T. C. Beers, "The Formation of the Heaviest Elements," Phys. Today **71**, 30–37 (2018). https://doi.org/10.1063/PT.3.3815

22 F. Thielmann, M. Eichler, I. Panov, B. Wehmeyer, "Neutron Star Mergers and Nucleosynthesis of Heavy Elements," Ann. Rev. Nuclear Particle Sci. **67**, 253–274 (2017). https://doi.org/10.1146/annurev-nucl-101916-123246

23 M.-H. Chen et al., "Neutron Star Mergers as the Dominant Contributor to the Production of Heavy r-process Elements," MNRAS **529**, 1154–1160 (2024). https://doi.org/10.1093/mnras/stae475

24 B. P. Abbott et al., "Multi-messenger Observations of a Binary Neutron Star Merger," ApJL **848**, L12 (59) (2017). https://doi.org/10.3847/2041-8213/aa91c9

25 W. Allen, "The UFO Menace," in *Side Effects* (Random House, New York, 1980)

26 A. S. Eddington, Observatory **58**, 38 (1935)

27 A. Eddington, *The Internal Constitution of the Stars* (Cambridge University Press, Cambridge, 1926)

28 J. Gribbin, *The Scientists: A History of Science Told through the Lives of Its Greatest Inventors* (Random House, New York, 2003)

4

From Dust to Us: Our Cosmic Connection

Summary The life cycle of stars establishes a profound interconnectedness across the Universe. From the birth of stars in distant galaxies to the formation of planets within our Solar System, this cosmic thread weaves through space and time, linking celestial bodies in an intricate dance of creation and destruction. As Joni Mitchell reminds us in her 1969 song "Woodstock," "we"—blood, bones, and brains, and all we see around us—"are stardust," children of the stars, all made of the same stuff ("billion-year-old carbon"), all brought into existence with the birth of the Universe and the birth and death of stars. All the chemical elements except for hydrogen and most of the helium—the oxygen and nitrogen in the air we breathe, the calcium in our bones and teeth, the iron in the hemoglobin in our blood, and so much more—have been built up over time by nuclear reactions within and among stars that eventually dispersed this processed stellar material, now comprising nearly all the chemical elements found on Earth, into the surrounding interstellar medium where, in time, new stars and planets—and, in at least one case, ponies and people—eventually formed. Stardust, the raw material for life, is spread throughout the Universe: life as we know it is an inevitable consequence of stellar evolution, the life cycle of the stars. Although they may not govern our conditions, the stars above us have certainly helped *create* our conditions—and our cosmic connection—and, significantly, the conditions for life throughout the Universe.

R. Fleck, *We Are Stardust*, https://doi.org/10.1007/978-3-031-67275-0_4

Twinkle, twinkle little star
How I've wondered what WE are.
Now I know you're made of dust
Now I know you're just like us.
Twinkle, Twinkle oh so far,
Now I know I am a star.

— Robert K Davis [1]

All of the rocky and metallic material we stand on, the iron in our blood, the
calcium in our teeth, the carbon in our genes were produced billions of years ago
in the interior of a red-giant star. We are made of star-stuff.

— Carl Sagan, *The Cosmic Connection* [2, pp. 189–190]

… the fixed stars … are not both lit and burnt out in the life of a man—yet they
too are his distant relations.

— American naturalist Henry David Thoreau writing at
Walden Pond (1846)

Indeed, the stars, "oh so far," are "dust … like us." "We are," as the astronomer Carl Sagan so famously proclaimed, "made of star-stuff." And while our timescales are grossly mismatched—"the fixed stars … are not both lit and burnt out in the life of a man" (recall Note 4 of Chap. 3)—the stars are, as Thoreau presciently perceived, our "distant relations." All of the hydrogen, most of the helium, and some of the lithium in the Universe were produced in the first few minutes after the Big Bang brought the Universe into existence 13.8 billion years ago.[1] The stars, either directly through slow hydrostatic or rapid explosive nucleosynthesis, or indirectly through cosmic-ray spallation (cosmic rays get their energy from high-energy stellar processes such as supernovae), gave us all the rest of the naturally occurring chemical elements, almost as an afterthought to satisfying their energy requirements. *This is our cosmic connection*, our direct link to distant events spread across the Universe in space and time reaching back to the beginning, back to the Big Bang, and continuing through the birth and death of successive generations of stars providing the raw material for new stars, planets, and, on at least one occasion, people. We are stardust (Fig. 4.1)—in a very real sense, children of the stars—"star folk" made from chemical elements ("star stuff") created by

---

[1] Think about that the next time you have a drink of water, H_2O, two of the three atoms of which are Big-Bang-made hydrogen. But then *all* matter ultimately derives from the protons, neutrons, and electrons produced by the Big Bang; the stars have only rearranged these atomic constituents into the wonderous array of diverse chemical elements in the Universe today. Even primordial hydrogen has to a greater or lesser extent been rearranged as electrons are driven away from their originally paired protons (in a process called *ionization*), only to recombine (*recombination*) with other protons to form subsequent generations of hydrogen. We and everything around us are all part of a Universe in a state of constant flux.

Fig. 4.1 The Stardust hotel and casino in Las Vegas, Nevada, sometime between 1968 and 1977. "For you are dust," the Bible reminds us (Genesis 3:19), "and to dust you shall return," just as this Las Vegas landmark eventually did. (*Wikimedia Commons, public domain*)

stellar nuclear processes and spread throughout the Galaxy during the various stages of stellar evolution. A Universe that would make element abundances different from ours—for example, more boron than carbon or more fluorine than oxygen, each pair being neighbors in the periodic table—would be very different from ours in very many ways, chemically, physically, and, if biology were even possible, biologically.

The building blocks of life, the Universe, and everything, synthesized in stellar processes, are found throughout the dusty component of interstellar material. More than 700 molecules have been discovered there or in circumstellar envelopes and planetary nebulae, many of which are organic in nature including glycine, the simplest amino acid, and various species of polycyclic aromatic hydrocarbons (PAHs) and molecules containing benzene rings [3]. These molecules provide important information on physical conditions and evolutionary stages of star-forming regions, but, most importantly for our story here, they provide the molecular seeds for life in the Universe.

In addition to this organic soup wafting through galaxies, interstellar dust grains—stardust—make up about one percent of the interstellar medium by mass. As mentioned in the previous chapter, these are hard, refractory, submicron-size agglomerates of silicates and carbonaceous material encased by icy mantles, rich in silicon carbide (SiC), graphite, aluminum oxide, and various metal oxide spinel minerals, among other such solids that condensed at high temperatures from a cooling gas such as found in the atmospheres of cool oxygen- and carbon-rich red giants, stellar winds, and supernova explosions. Isotopic ratios in these grains have, not surprisingly, provided important information on the many particulars of stellar nucleosynthesis. For

Fig. 4.2 Artist sketch of the protosolar nebula—the early Solar System—showing planets forming in a dusty protoplanetary disk encircling a young, forming Sun. (*Courtesy of NASA, public domain*)

example, the heavy elements within the silicon carbide grains are nearly pure s-process isotopes formed within red-giant stellar winds; these stars are the main source of s-process nucleosynthetic elements and they have atmospheres observed to be highly enriched in dredged-up s-process elements.

As outlined in the previous chapter, stars form in cold, dense, dusty interstellar clouds enriched with heavy elements produced and dispersed by previous generations of stars. Our star, the Sun, formed nearly 5 billion years ago from a cloud of gas and dust enriched to a metal content of 2%, matching the metal abundance in the solar atmosphere today (recall that astronomers consider all elements heavier than hydrogen and helium to be metals). The chaotic, turbulent nature of the interstellar medium endows these clouds with rotation that increases as they collapse due to conservation of angular momentum (the same principle that enables an ice skater to spin rapidly when "collapsing" her arms close to her chest). Eventually, centrifugal forces halt contraction in the equatorial plane perpendicular to the rotation axis, causing the collapsing nebula to leave behind a protoplanetary disk from which the planets, including Earth, formed (Fig. 4.2 and recall Fig. 3.5b); *planet formation, we now know, is a natural byproduct of star formation.* Seventy percent of stars in the Galaxy are M dwarfs—the "silent majority" because they are so faint—and 30% of those observed thus far host planets, more than any other spectral type.

Earth alone, of all the planets in our Solar System, developed in the Goldilocks-like *habitable zone*, not too close to the Sun for water to boil away (as occurred on Mercury and Venus), and not too far for water to freeze to ice (as occurred in the outer Solar System), a perfect environment—indeed, a quintessential requirement—for life as we know it. Remaining in the liquid state over this relatively large range in temperature makes water the medium of choice as a solvent to facilitate biochemical reactions and nutrient transport; hence NASA's "follow the water" strategy in the search for life and habitable environments. Analysis of meteorites, pristine pieces of asteroidal and cometary debris, recovered today indicates that asteroids, chunks of space rock and metal, and comets, "dirty snowballs" composed of various ices and other volatiles as well as silicate minerals and prebiotic organic compounds, delivered water and life's building blocks to our young planet during the Late Heavy Bombardment period some 3.8–4.1 billion years ago. (Thousands of tons of cosmic dust continue to rain down on Earth each year—nothing compared to the hundreds of millions of tons of plastic waste that foul our home annually.) Some seventy amino acids were found in the Murchison meteorite that landed near the Australian town of that name in 1969, several of which are among the twenty used by life on Earth, as were silicon carbide inclusions that are seven billion years old, 2.5 billion years older than Earth. As the astrobiologist and planetary scientist Christopher McKay noted, "The chemistry of life appears to be part of cosmic organic chemistry" [4, p. 195].

In a process not yet completely understood or replicated in the laboratory, life originated abiogenetically from pre-biotic organic material, likely in warm, tidal pools (Fig. 4.3) or perhaps along hydrothermal vents on the ocean floor. Meteorites, interplanetary dust, and perhaps organic hazes in Earth's early methane-rich atmosphere, a world compositionally akin to Saturn's moon Titan today, contained a vast inventory of life's building blocks including nucleobases that make up ribonucleic acid (RNA) as well as the amino acids that make up proteins, and thus may have seeded warm little ponds for life [5]. The early experiments [6] by the American chemists Harold Urey (1893–1981) and Stanley Miller (1930–2007), replicated more recently by several other groups, confirmed the proposal [7] by Russian biochemist Aleksandr Oparin (1894–1980) that the conditions on primitive Earth favored chemical reactions that synthesized complex organic compounds including amino acids from simpler inorganic precursors, a concept known as the "chemical evolution" theory of the origin of life. "This transition from the inanimate to the living was an almost incredible sequence of highly improbable events," the American astronomer Harlow Shapley admitted [8, p. 341],

> but in a universe containing more than 10^{20} stars capable of building planetary systems and over a period of at least 10^{10} earth-years, almost anything can happen repeatedly. And obviously this unlikely sequence did happen at least once in this solar system, for here we are, the timid descendants of some rather nauseating gases and sundry flashes of lightning!

Fig. 4.3 Many biologists believe that life on Earth began in a primordial soup of pre-biotic organic compounds nearly four billion years ago. You won't find this soup on your grocer's shelves. (*Photograph by the author from author's collection*)

One thing is certain: *all life forms on Earth share a common carbon chemistry and a common genetic code, suggesting a common chemical origin for life.*[2] And as soon as the conditions necessary for life as we know it were in place, not long after the Late Heavy Bombardment period, life arose, astoundingly adaptable and remarkably resilient, a wondrously diverse tapestry woven— for it is the pattern, not the cloth, that defines life (along with agency and purpose, according to author Phillip Ball [10])—from so simple a thing as the carbon atom which has the ability to form the strong and multiple bonds required for the construction of the complex macromolecules necessary to

---

[2] Alternative life chemistries, a staple of science fiction, are, of course, possible. I am reminded of *Star Trek*'s Horta, a highly intelligent rock-like life form based not on carbon but on silicon, a major component of rock and an element having an outer electron configuration matching carbon's and hence chemically similar to carbon (it's in the same "family"—vertical column—in the periodic table (recall Fig. 3.14); but then so is tin (Sn): whence the Tin Man from *The Wizard of Oz*?). And what kind of chemistry forms the basis of the bizarre life forms on author Robert Forward's fictional neutron star, Dragon's Egg [9]? Of course, alien life could be so alien that we won't recognize it, as was the case with the "Black Cloud" in the 1957 sci-fi novel of the same name by the astronomer Fred Hoyle (the Hoyle of B[2]FH introduced earlier), an idea resurrected by the "Cloud Creature," a living, intelligent interstellar cloud featured in the TV series *Star Trek*, with whom Spock performed a Vulcan mind meld to learn its intentions.

carry out life processes at the cellular level.[3] The rate of variation in RNA implies a common ancestor for *all* life about 3.5 billion years ago. As the final words of Charles Darwin's magisterial *Origin of Species*, published in 1859, proclaim, "from so simple a beginning endless forms most beautiful and most wonderful have been, and are being evolved."

Besides being major producers and distributors of the chemical elements, expanding supernova shock waves trigger the formation of new stars and are the main source from outside the Solar System of cosmic rays that cause many of the mutations that drive the evolution of life on Earth. On the other hand, a kilonova—an explosion caused by two colliding neutron stars (discussed in Chap. 3)—located within a few dozen light years of Earth could bathe the planet in cosmic rays and exterminate life for thousands of years. A gamma ray burst (GRB; recall Chap. 3) can also have harmful and destructive effects on life. Fortunately, all GRBs observed to date have occurred well outside our Galaxy and have been harmless to Earth. However, if a GRB were to occur within several thousand light years, and its emission were beamed straight towards Earth, the effects could be harmful and potentially devastating for terrestrial ecosystems. Life is at once resilient—witness the *extremophiles* (from the Latin and Greek meaning "love of extremes") living under miles of rock and in the frozen Antarctic and boiling hot springs of Yellowstone National Park—and fragile.

Not surprisingly, life is built from the most abundant raw materials available. The human body contains detectable traces of some 60 chemical elements, about half of which are thought to be necessary for life and health: you can thank your lucky stars for that. CHON—*C*arbon, *H*ydrogen, *O*xygen, and *N*itrogen—the four most common reactive elements in the Universe by number (helium being inert) as a result of their importance in the early stages of stellar nucleosynthesis (recall from Chap. 3 the CNO cycle, for example) are the four most abundant chemical elements in the human body, accounting for 96% of the total (Fig. 4.4). Together, calcium and phosphorus make up a little more than half the rest, with somewhat less than 1% composed of the five essential elements: potassium (K), sulfur, sodium, chlorine (Cl), and magnesium. All 11 of these most abundant elements are

---

[3] Recall from Note 20 in Chapter 3 that carbon is special also at the nuclear level: a resonant energy state in the carbon-12 nucleus ensures that a small percentage of triple-alpha reactions in red giant stars produce carbon rather than decaying back to the original alpha particles, a fine tuning to "just so" conditions that, if it didn't exist, neither would we–or anything else in the Universe made of elements heavier than those cooked in the early Universe. Thus it is, for more reasons than one, that, as the British physicist and mathematician, Sir James Jeans, proclaimed in his 1930 *The Mysterious Universe* [11, p. 8], "Life exists in the Universe only because the carbon atom possesses certain exceptional properties."

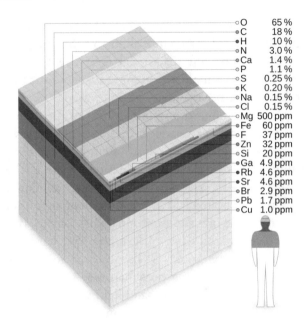

○O	65 %
●C	18 %
●H	10 %
○N	3.0 %
●Ca	1.4 %
○P	1.1 %
○S	0.25 %
●K	0.20 %
○Na	0.15 %
○Cl	0.15 %
○Mg	500 ppm
●Fe	60 ppm
○F	37 ppm
●Zn	32 ppm
○Si	20 ppm
●Ga	4.9 ppm
●Rb	4.6 ppm
●Sr	4.6 ppm
●Br	2.9 ppm
○Pb	1.7 ppm
○Cu	1.0 ppm

Fig. 4.4 Parts-per-million cube showing the relative abundance by mass of elements in an average adult human down to 1 ppm (people with more fat will have a higher proportion of carbon). Similar percentages occur in the common gut bacterium *Escherichia coli* (*E. coli*) which however, being boneless, contains much less calcium. (*Wikimedia Commons,* https://commons.wikimedia.org/wiki/User:Cmglee, https://creativecommons.org/licenses/by-sa/4.0/)

necessary for life, along with more than a dozen "trace" elements which together, at about 10 g for a typical body weight, add up to less than the body mass of magnesium, the least common of the 11 non-trace elements. Less common elements, including the iron in your blood (note the nearly factor of ten drop in abundance after magnesium), along with manganese (Mn), cobalt, copper (Cu), zinc (Zn), selenium (Se), molybdenum (Mo), and iodine (I), are also essential to life and can be found on the label of any good brand of multivitamins and minerals.

The two most common elements, hydrogen and oxygen, combine to form life's sine qua non molecule: water, a compound making up more than half our weight, more oxygen by mass but more hydrogen by number. Magnetic resonance imaging (MRI) utilizes the magnetic properties of protons in the hydrogen that is so abundant in bodily water and fat. The next most abundant element, carbon, is the basis of organic chemistry, the chemistry of life (recall Note 6 of Chap. 1), and nitrogen is an essential element in the five nitrogenous bases—adenine (A), cytosine (C), guanine (G), thymine (T), and uracil (U)—of the nucleic acids RNA (*Ribo*N*ucleic* A*cid*) and DNA

(*DeoxyriboNucleic Acid*) that carry the genetic code (with the bases A, G, C, and T being found in DNA while A, G, C, and U are found in RNA) and of the 20 amino acids from which are built all proteins, the basic building blocks of biology.

Not all elements found in the human body play a role in life. Indeed, some, such as cesium and titanium, two of the so-called "heavy metals," are common contaminants without function, while others, such as arsenic (As), cadmium (Cd), mercury (Hg), and lead (Pb), are active toxins. Mammals, in general, require smaller amounts of many other elements such as silicon, boron, nickel, and vanadium (V). Bromine (Br) is an essential element for some bacteria, fungi, diatoms, and seaweeds, and is used opportunistically in eosinophils, a type of white blood cell that supports the immune system in humans.

Most of the elements necessary for life are relatively common in Earth's crust which, due to various physical and chemical processes during its formation so close to the forming Sun, differs significantly from the cosmic abundance of elements; the "gas" giant planets Jupiter and Saturn, however, are similar in composition to the Sun. As Fig. 4.5 illustrates, Earth is rich in high-temperature refractory materials such as iron and silicates, and lacking in volatile materials (as is our Moon which formed after a Mars-size planetesimal slammed into Earth during its formation). The elements aluminum (Al) and silicon, although very common in Earth's crust, are conspicuously rare in the human body (if not in *Star Trek*'s Horta, a sentient non-humanoid silicon life form native to the fictional planet Janus VI; recall Note 2); large amounts of Al can be toxic in humans. The four most abundant elements, together making up 93% of Earth's mass, are iron (32%), the main ingredient of Earth's metallic core, oxygen (30%) and silicon (16%), combining to form the silicate minerals—rocks—that make up the bulk of Earth's mantle and crust, and magnesium (15%). Carbon, so important to life, makes up only 0.073% of our home planet, and hydrogen, the most abundant element in the Universe, only 0.026%. The young Earth was entirely molten due to the heat generated by impacting material during its formation, allowing heavier metals to sink and form a core while the lightest minerals floated to the surface like so much scum to form the crust. This differentiated (layered) structure was hinted at more than two centuries ago when Earth's bulk density was found to be about 5.5 g/cm^3 (5.5 times the density of water), about twice that of crustal rock and half that of its iron-nickel core (note from Fig. 4.5 that nickel is fifth in abundance).[4]

---

[4] The *bulk density* of an object—its mass divided by its volume—is a powerful indicator of its composition and is an easy quantity to calculate once the object's mass and size (and hence volume)

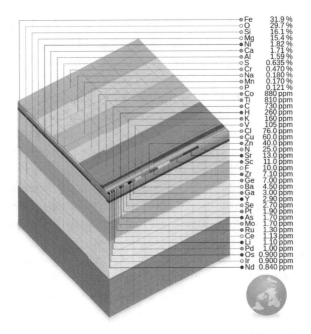

Fe	31.9 %	
O	29.7 %	
Si	16.1 %	
Mg	15.4 %	
Ni	1.82 %	
Ca	1.71 %	
Al	1.59 %	
S	0.635 %	
Cr	0.470 %	
Na	0.180 %	
Mn	0.170 %	
P	0.121 %	
Co	880 ppm	
Ti	810 ppm	
C	730 ppm	
H	260 ppm	
K	160 ppm	
V	105 ppm	
Cl	76.0 ppm	
Cu	60.0 ppm	
Zn	40.0 ppm	
N	25.0 ppm	
Sr	13.0 ppm	
Sc	11.0 ppm	
F	10.0 ppm	
Zr	7.10 ppm	
Ge	7.00 ppm	
Ba	4.50 ppm	
Ga	3.00 ppm	
Y	2.90 ppm	
Se	2.70 ppm	
Pt	1.90 ppm	
As	1.70 ppm	
Mo	1.70 ppm	
Ru	1.30 ppm	
Ce	1.13 ppm	
Li	1.10 ppm	
Pd	1.00 ppm	
Os	0.900 ppm	
Ir	0.900 ppm	
Nd	0.840 ppm	

Fig. 4.5 Parts-per-million cube showing the relative abundance by mass of elements of the bulk Earth down to around 1 ppm. (*Wikimedia Commons,* https://commons. wikimedia.org/wiki/User:Cmglee, https://creativecommons.org/licenses/by-sa/4.0/)

We've already examined two of the three "big bangs" discussed by Dauber and Muller in their book *The Three Big Bangs: Comet Crashes, Exploding Stars, and the Creation of the Universe* [12], those being exploding stars and the biggest, if quietest, of big bangs, *the* Big Bang that resulted in the creation of the Universe. The remaining and most recent bang—a comet/asteroid

---

are known. You can do a lot of planetary science knowing only the bulk density of solar-system objects. For example, the bulk density—and hence bulk composition—of the planet Venus is similar to Earth's, while that of Mars and the Moon is somewhat less due to their lower metal abundance; Mercury, closest to the Sun, is densest of all due to its substantial metallic core. Jupiter and Saturn, being of solar composition and thus made mainly of the light elements hydrogen and helium, have bulk densities similar to the Sun, Jupiter and the Sun being slightly denser than water, and Saturn being the only planet less dense than water (which means it would float if you could find a bathtub big enough to put it in—giving new meaning to its label "Lord of the Rings"). The bulk density of many of the major satellites of the outer planets, like that of the dwarf planet Pluto, is midway between the density of ice and rock, and thus contain more or less equal amounts of ice and rock; the decreasing density of the four major (Galilean) satellites of Jupiter with increasing distance from the planet closely mirrors the density-(solar)distance trend exhibited by the four inner earth-like planets. Looking at the minor members of the Solar System—minor is size but major in importance due to their pristine, primitive state—the bulk density of asteroids tells us they are made of rock and/or metal, while that of comets suggests a fluffy ice and rock composition. A lot of information from such a simple and easily measured quantity.

Fig. 4.6 Casts of a hypothetical reptilian humanoid in front of a *T. Rex* at the Smithsonian Museum of Natural History, Washington DC, suggesting what "we" may have looked like today had the asteroid missed Earth and the Age of Reptiles continued. In the original *Star Trek* TV series, Captain Kirk fights an extraterrestrial reptilian Gorn, evidently from a planet that escaped asteroid bombardment and subsequent short-circuiting of reptilian evolution. Dozens of impact craters have been identified on Earth, and every solid surface in the Solar System, from our Moon to distant Pluto, is peppered with them. (*Photograph by the author*)

crash—is another cosmic event that impacts, so to speak, our cosmic connection, if only in the sense of directing the evolution of life on Earth towards our sometimes rational, thinking selves, *Homo sapiens* (Fig. 4.6). Indeed, so far as we know, we are the only sentient creatures in the Cosmos who are aware of the Universe and who understand how—if not why—we got here and how we are connected to the Cosmos. As Carl Sagan comments in *Cosmos* [13, p. 21], "until we find more intelligent beings elsewhere, we are ourselves the most spectacular [if not most dangerous, I would add] of all the transformations [of matter and energy]—the remote descendants of the Big Bang, dedicated to understanding and further transforming the Cosmos from which we spring." Sagan argues that we therefore have a "cosmic responsibility" to learn more about the Cosmos and our place in it. I agree.

* * *

And so it goes, from dust to us—and so much more, including rust (iron oxide, with both iron and oxygen coming from the stars). Astrobiologists—those who study life in the Universe—debate the presence of life, now or in the past, on Mars and in the subsurface oceans thought to exist on Jupiter's moon Europa and perhaps Saturn's Enceladus, that may force us to revisit our definition of a star's habitable zone which may be much larger than currently conceived. The discovery at the end of the last century of planets around other stars, another highlight of twentieth-century science (thousands of these exoplanets are now known, some of which exhibit surface and atmospheric biosignatures), extends the province of planetary science and astrobiology beyond our Solar System out into the Universe at large [14, 15]. The search continues.

And with this in mind, it is perhaps fitting to recall the nineteenth-century Scottish essayist Thomas Carlyle's take on the possibility of extraterrestrial life on other worlds (he who admitted "I don't pretend to understand the Universe—it's a great deal bigger than I am"): "A sad spectacle. If they be inhabited, what a scope for misery and folly. If they be not inhabited, what a waste of space." A waste of space, indeed: the odds that we are alone in the Universe—that life here is merely a "happy accident"—are pretty slim given the vastness of space and the ubiquity of stardust. In his aptly titled 1885 poem "Vastness," the Victorian Poet Laureate Alfred, Lord Tennyson, enthralled by astronomy all his life, wondered whether "Many a planet by many a Sun may roll with the dust of a vanished race." Two centuries earlier, the Dutch scientist Christiaan Huygens asked the readers of his popular *Cosmotheoros*, translated and published posthumously in 1698 under the title *The Celestial Worlds Discover'd: or, Conjectures Concerning the Inhabitants, Plants and Productions of the Worlds in the Planets*, if

> Earth [be] one of the Planets of equal dignity and honor with the rest, who would venture to say, that nowhere else were to be found any that enjoy'd the glorious sight of Nature's Opera? Or if there were any Fellow-Spectators, yet we were the only ones that had dived deep to the secrets and knowledge of it?

Huygens reasoned that since Earth held no privileged position in the Copernican universe, it was unreasonable to suppose that life should be restricted to Earth alone, a prospect that would have given Earth, once again, a privileged position.

Earlier in the seventeenth century, the English metaphysical poet John Milton speculated in his epic work *Paradise Lost* on the habitability of the innumerable stars revealed by the newly invented telescope:

> … stars
> Numerous, and every star perhaps a world
> Of destined habitation.

Similarly, in eighteenth-century England, Alexander Pope wondered in his *Essays on Man*, as the Neapolitan Dominican monk Giordano Bruno did (before being burned at the stake in 1600 for this and other heresies),

> What other planets circle other suns,
> What vary'd Being peoples ev'ry star….

In the nineteenth century, the American writer-physician Oliver Wendell Holmes asked in *The Secret of the Stars*:

> Are all these worlds, that speed their circling flight,
> Dumb, vacant, soulless—baubles of the night?
> … Or rolls a sphere in each expanding zone
> Crowned with a life as varied as our own?

More recently, author John Gribbin concludes his book on *Stardust* with this important reminder: "We are made of stardust because we are a natural consequence of the existence of stars, and from this perspective it is impossible to believe that we are alone in the Universe" [16, p. 187]. Indeed, much like plant seeds—and sometimes life itself—that waft or raft across miles of ocean to populate isolated islands, it has been suggested that life exists throughout the Universe and has seeded life here on Earth and elsewhere across the Galaxy, an idea known as *panspermia* (from the Greek meaning "seeds everywhere").[5] *The answer to the question "Are we alone?" addresses our uniqueness in the Universe and is thus of the utmost importance. It calibrates our very existence.* It would be very non-Copernican of us to claim that we are alone and therefore special in this immense and possibly infinite Cosmos.

Concerning the possibility of *intelligent* life in the Universe (whatever "intelligent" means: one often wonders whether there is intelligent life here

---

[5] In an interesting—if decidedly disconcerting—spin on panspermia, the astronomer Thomas Gold, whom we met in Chapter 2 as a co-conspirator of steady-state cosmology and, again, in Chap. 3 as the architect of our understanding of pulsars as rapidly rotating neutron stars, proposed that life on Earth originated from cosmic garbage—leftovers left behind by picnicking aliens [17, p. 65]. Evidently, they missed the "Please Take Your Trash with You" sign.

on Earth, what with our technology outpacing our wisdom), I side with Bill Watterson's *Calvin and Hobbes* (8 November 1989): "the surest sign that intelligent life exists elsewhere in the Universe is that it has never tried to contact us." In any case, the words penned by the sage Sagan half a century ago as I write mine ring true, even more so today: "Extraterrestrial life is an idea whose time has come" [2, p. viii]. The quest to understand the origin of life on Earth—yet another "origins" mystery (recall Note 2 of Chap. 2)—and to explore the prospects for life elsewhere in the Universe must count as one the most profound of human endeavors (recall Fig. 1.1).

> He, who thro' vast immensity can pierce,
> See worlds on worlds compose one Universe.
> — Alexander Pope, "Essay on Man" (1733-34; Epistole I, lines 23–24)

References

1 Wikipedia, "Twinkle, twinkle, little star." https://bcn.boulder.co.us/~neal/poetry/twinkle.html. Accessed 12 May 2024

2 C. Sagan, *Carl Sagan's Cosmic Connection: An Extraterrestrial Perspective* (Cambridge University Press, 2000; orig. publ. *The cosmic connection*, Doubleday & Company Inc., 1973)

3 T. J. Millar, C. Walsh, M. Van de Sande, A. J. Markwick, "The UMIST Database for Astrochemistry 2022," Astron. Astrophys. **682**, A109 (2024). https://doi.org/10.1051/0004-6361/202346908

4 C. P. McKay, "The Origin and Evolution of Life in the Universe," in *Origin and Evolution of the Universe: From Big Bang to Exobiology*, ed. by M. A. Malkan, B. Zuckerman, 2nd edn (World Scientific, Singapore, 2020)

5 B. K. D. Pearce, S. M. Hörst, J. A. Sebree, C. He, "Organic Hazes as a Source of Life's Building Blocks to Warm Little Ponds on the Hadean Earth," Planet. Sci. J. **5**, 23, 20pp (2024). https://doi.org/10.3847/PSJ/ad17bd

6 S. L. Miller, "The Formation of Organic Compounds on the Primitive Earth," Ann. NY Acad. Sci. **69**, 260–275 (1957). https://doi.org/10.1111/j.1749-6632.1957.tb49662.x

7 A.I. Oparin, *The Origin of Life* (MacMillan Co, New York, 1938)

8 J. Silk, *The Big Bang*, revised and updated edn. (W. H. Freeman & Co., New York, 1989)

9 R. Forward, *Dragon's Egg* (Del Ray, New York, 1980)

10 P. Ball, *How Life Works: a User's Guide to the New Biology* (University of Chicago Press, Chicago, 2023)

11 J. Jeans, *The Mysterious Universe* (Macmillan Co, New York, 1930)

12 P. M. Dauber, R.A. Muller, *The Three Big Bangs: Comet Crashes, Exploding Stars, and the Creation of the Universe* (Addison-Wesley Pub Co, Boston, 1995)

13 C. Sagan, *Cosmos* (Random House, New York, 1980)

14 C. Impey, *The Living Cosmos: Our Search for Life in the Universe* (Random House, New York, 2007)

15 J. Bennett, S. Shostak, N. Schneider, M. MacGregor, *Life in the Universe*, 5th edn. (Princeton University Press, Princeton and Oxford, 2022)

16 J. Gribbin, *Stardust* (The Penguin Group, Chatham, UK, 2000)

17 T. Gold, "Cosmic Garbage," Air Force and Space Diges **43** (1960)

5

Final Thoughts

Summary The title of a monumental painting in the Boston Museum of Fine Arts by the French post-Impressionist artist Paul Gauguin, his 1897 *D'où Venons Nous / Que Sommes Nous / Où Allons Nous* (*Where do we come from? What are we? Where are we going?*), appears in French verse in the upper-left corner of the painting which was made during the artist's Polynesian period. Ever since we became conscious of ourselves and the world around us, we have asked "ultimate" questions such as these regarding our origin, purpose, and destiny. It is significant that when Gauguin posed them, science was not in a position to provide answers; one would have had to consult philosophers or theologians. Today, however, more than a century later, science has come a long way in answering these most profound of questions that still lie at the heart of philosophy and modern science—even if, to the dismay of author John Gribbin, "hardly anybody outside a small circle of scientific specialists seems to have noticed" that "[w]e have answered the biggest question of them all—where do we come from?"—Gauguin's leading question here. We now know that *we are stardust, children of the stars*, one of the most profound and inspiring discoveries ever made, certainly *the* most fundamental and fascinating finding about ourselves and our connection to the Cosmos. Understanding our cosmic connection—that the stuff we are made of traces its origin to nuclear processes accompanying the Big Bang and thereafter to billions of years of the birth and death of generation after generation of stars—is an important and beautiful story that deserves more attention.

We are made of stardust. Every atom of every element in your body except for hydrogen has been manufactured inside stars, scattered across the Universe in great stellar explosions, and recycled to become part of you.

– John Gribbin [1, p. 1]

… we are the local embodiment of a Cosmos grown to self-awareness. We have begun to contemplate our origins: starstuff pondering the stars, organized assemblages of ten billion billion billion atoms; tracing the long journey by which, here at least, consciousness arose…. Our obligation to survive is owed not just to ourselves but also to that Cosmos, ancient and vast, from which we sprang.

– Carl Sagan [2, p. 345]

The highest wisdom has but one science, the science of the whole, the science explaining the Creation and man's place in it.

– Russian author Leo Tolstoy, *War and Peace* (1869)

Since our earliest times on this planet, we have wondered about the world around us and about our place and purpose in it. We wanted to count in the grand scheme of things, and longed to be connected to the Cosmos. We fashioned stories—myths—many of which expounded some aspect of cosmic order connecting our human microcosm to the cosmic macrocosm (Fig. 5.1; recall also Figs. 1.2 and 2.1 and Note 22 of Chap. 2).[1] And we still do. Sometimes in our art (as in Figs. 1.2, 2.1, and 5.1), sometimes in our science (recall Figs. 2.16, 2.20, and 3.13). There is something comforting and uplifting about feeling at home in the Universe—a *yūgen*, the Japanese would say, an awareness of the Universe that triggers a mysterious and sublime profundity too deep and too powerful for words.

One of the earliest of several works of art through history portraying a cosmic connection between the celestial macrocosm and the terrestrial/human microcosm, the oldest artifact with possible scientific underpinnings, and the earliest evidence we have that our ancestors may have recognized a resonance between celestial and human cycles, is the *Venus of Laussel* (recall Fig. 5.1a), a 46-cm-high bas-relief fertility figure from the Upper Paleolithic

---

[1] Modern physics has revealed the atom as a sort of miniature Solar System, with most of its mass concentrated in a central nucleus, about which move tiny electrons like so many planets, and, as reviewed in Chap. 2, modern cosmology shows that our understanding of the largest scales imaginable in the Universe is strongly coupled to our understanding of the smallest subatomic particles in nature, tying the physics of the very large with that of the very small, just as this same physics of the very small—quantum physics—explains the workings of galaxies of stars like our Sun, the largest object we can discern with our eyes. The Cosmos abounds in microcosm-macrocosm connections, with our understanding of it linked across all scales. Fracture, to take one more example—the process by which external forces break up a solid object—provides compelling evidence for the existence of atoms and thus links the macroscopic and microscopic worlds in a decidedly mechanical way [3]. As the nineteenth-century Suquamish Indian Chief Seattle apocryphally proclaimed, "All things are connected."

Fig. 5.1 **a** The *Venus of Laussel*, a fertility figure from the Upper Paleolithic, the earliest evidence we have that our ancestors may have recognized a resonance between celestial (macrocosm) and human (microcosm) cycles. (*Wikimedia Commons, user 120*, https://creativecommons.org/licenses/by/3.0/deed.en). **b** Baboons, having a reputation like roosters for being active and noisy at dawn, observe the rising Sun being rolled across a star-studded sky by a scarab beetle, a representation of the solar deity Khepri, god of creation and rebirth, in this exquisitely fashioned pendant found in the tomb of Tutankhamun ("King Tut"), now in the Egyptian Museum in Cairo. An early example of microcosm-macrocosm analogy with the small mimicking the large, the scarab beetle is a manifestation of the Sun god Ra rolling the (macro) Sun before him like a ball of (micro) dung. The analogy runs deeper than a shared roundness of shape. The dung beetle lays its eggs in the rolled-up dung, thereby providing her young with their first meal when they enter the world. To the observant ancient Egyptian, unaware of our modern understanding of procreation, it would have appeared that life originated spontaneously from the dung. Just as the dung appeared to give life to beetle progeny in the microcosm, the Sun was revered for promoting life in the macrocosm. Interestingly, scientists have recently discovered that certain species of nocturnal scarabs navigate by moonlight or by the light of the Milky Way, while day-rolling species look to the Sun as their guiding star, amazingly direct ties between the microcosm and the macrocosm. (*Photograph by the author*)

(*paleo*, old; *lithic*, stone: hence Old Stone Age) discovered in the Dordogne department of southwest France, now residing in the Musée d'Aquitaine in Bordeaux. Her large, pendulous breasts, exaggerated hips, and left hand placed so delicately over her slightly swollen womb, clearly identify this as a "Venus" fertility figure, a motif common to this early period of art often found as freestanding figurines such as the famous *Venus of Willendorf*. In her right hand she holds a bison's horn shaped like a crescent Moon, both symbols of fertility. The mythic association of the Moon with fertility is well known and probably originated with our early recognition of the correlation between the Moon's cycle of phases and a woman's menstrual cycle, each lasting approximately one month. (The words Moon, month, and menstrual derive from a common Indo-European root.) The cycle of life and death—and of life after death—is mirrored in the cycle of lunar phases. It has been suggested that the thirteen notches in the bison horn may represent the approximate number of lunar cycles—months or "moons"—in the annual solar cycle, or may perhaps mark the number of days taken by a reborn dark "new" Moon to reach its fully illuminated phase. Faith in the rule of celestial bodies over our lives was not unreasonable given the cyclical governance evident in all aspects of the seasonal year. That the Moon exerts a direct influence on Earth, gravitationally controlling the ebb and flow of the tides, would be discovered much later in history.[2]

Mindful that with the *Venus of Laussel* we are dealing with a prehistoric, preliterate society, which therefore left us no record of what exactly they had in mind, with this interpretation this interesting artifact becomes *the earliest surviving evidence for the assignment of astrological influences on the human condition, and the concomitant acknowledgment of fundamental sympathies and correspondences—cosmic cues—between the terrestrial/human microcosm* ("little world") *and the celestial macrocosm* ("large world")—a *microcosm-macrocosm analogy*—whereby the small mimics in miniature the large: "as above, so below," an idea that resurfaces throughout the history of art and science. More importantly, the *Venus of Laussel* becomes *an early record of the human attempt to observe and interpret the workings of the world we are a part of, revealing a bold self-confidence in our ability to understand the world around us and the prospect of a connection—a resonance—between the way we think and the way the world works, the very prerequisites for a scientific understanding of the world*. It is not at all surprising that one of the earliest pieces of representational art reflects our earliest attempts to understand the workings of nature: art and science were coeval creations of our passionately curious and creative ancestors. As Albert Einstein once remarked [4, p. 130]: "The most beautiful experience we can have is the mysterious. It is the fundamental emotion which stands at the cradle of true art and true science." Whatever other emotions they may have shared, the earliest humans must have stood in awe at the myriad of mysteries around them.

We began our journey exploring our cosmic connection with French artist Paul Gauguin's monumental 1897 painting *D'où Venons Nous / Que Sommes*

---

[2] The Moon directly influences the fertility cycle of many aquatic species, including sea urchins in the Red Sea that spawn at full Moon, and coral over vast stretches of Australia's Great Barrier Reef that release trillions of eggs and sperm at precisely choreographed lunar moments.

Nous / Où Allons Nous (*Where do we come from? What are we? Where are we going?*), a title posing profound questions regarding our origin, purpose, and destiny (recall Fig. 1.1). We now know that we—and all we see around us—are intimately connected to the Cosmos, that we are a product of the workings and wonders of the Universe. We are stardust. We are, in a very real sense, children of the stars, star folk made from chemical elements ("star stuff") created by stellar processes and distributed throughout the Universe during the various stages of stellar evolution.

"We are just an advanced breed of monkeys on a minor planet of a very average star," the late cosmologist Stephen Hawking told a reporter for the 17 October 1988 issue of *Der Spiegel*, a German weekly news magazine. "But," he emphasized, "we can understand the universe. That makes us something very special." Physicist Jim Al-Khalili agrees: "… if we ever stop being curious about the universe and investigating how it—and we—came to be, then that is when we stop being human" [5, p. 281, the penultimate paragraph of his book]. Even the Irish playwright George Bernard Shaw appreciated the importance of carefully contemplating the meaning of the Universe [6, p. 650]:

> A person may be supremely able as a mathematician, engineer, parliamentary tactician or racing bookmaker; but if that person has contemplated the universe all through life without ever asking "What the devil does it all mean?" he (or she) is one of those people for whom Calvin accounted by placing them in his category of the predestinately damned.[3]

In his 1957 classic, *The Immense Journey*, the American anthropologist and naturalist Loren Eiseley bemoans the "burden of consciousness" we humans bear [7, pp. 161–162]:

> As … perhaps the only thinking animals in the entire sidereal universe—the burden of consciousness has grown heavy upon us. We watch the stars, but the signs are uncertain. We uncover the bones of the past and seek for our origins. There is a path there, but it appears to wander. The vagaries of the road may have a meaning, however; it is thus we torture ourselves.

---

[3] Compare this playwright's appreciation of searching for meaning in the Universe with the very different attitude of another playwright, the Englishman William Congreve, who in his 1695 *Love for Love* (Act 2, Scene 2) has this to say about the worth of the stars:
  If the Sun shine by Day, and the Stars by Night,
  why, we shall know one another's Faces without the help of a Candle,
  and that's all the Stars are good for.

In conclusion, we turn full circle back to where we began with more inspirational words and wisdom from the late astronomer Carl Sagan concerning our lonely planet's place in the Cosmos, these from his book *Pale Blue Dot: A Vision of the Human Future in Space* which was inspired by a photograph of Earth (Fig. 5.2; here occupying just one pixel) taken 14 February 1990 by NASA's Voyager 1 from beyond Neptune at a distance of nearly 4 billion miles from the Sun [8, pp. 6–7]:

… Look again at that dot. That's here. That's home. That's us. On it everyone you love, everyone you know, everyone you ever heard of, every human being who ever was, lived out their lives. The aggregate of our joy and suffering, thousands of confident religions, ideologies, and economic doctrines, every hunter and forager, every hero and coward, every creator and destroyer of civilizations, every king and peasant, every young couple in love, every mother and father, hopeful child, inventor and explorer, every teacher of morals, every corrupt politician, every "superstar," every "supreme leader," every saint and sinner in the history of our species lived there—on a mote of dust suspended in a sunbeam.

The earth is a very small stage in a vast cosmic arena. Think of the rivers of blood spilled by all those generals and emperors so that, in glory and in triumph, they could become the momentary masters of a fraction of a dot. Think of the endless cruelties visited by the inhabitants of one corner of this pixel on the scarcely distinguishable inhabitants of some other corner, how frequent their misunderstandings, how eager they are to kill one another, how fervent their hatreds.

Our posturings, our imagined self-importance, the delusion that we have some privileged position in the Universe, are challenged by this point of pale light. Our planet is a lonely speck in the great enveloping cosmic dark. In our obscurity—in all this vastness—there is no hint that help will come from elsewhere to save us from ourselves ….

It has been said that astronomy is a humbling and character building experience. There is perhaps no better demonstration of the folly of human conceits than this distant image of our tiny world. To me, it underscores our responsibility to deal more kindly and compassionately with one another, and to preserve and cherish that pale blue dot, the only home we've ever known.

Earlier, in his bestselling book *Cosmos* [2, p. 318], Sagan offers similar warnings:

National boundaries are not evident when we view the Earth from space. Fanatical ethnic or religious or national chauvinisms are a little difficult to maintain when we see our planet as a fragile blue crescent fading to become an inconspicuous point of light against the bastion and citadel of the stars.

Fig. 5.2 The "Pale Blue Dot," planet Earth (*circled*) as seen by the Voyager 1 space-craft on 14 Feb 1990, nearly 4 billion miles from home. (*Photograph courtesy of NASA, public domain*)

All the more reason to treat our home—and each other—with respect. "It's a message echoed by astronauts, who return from their cosmic travel urging us to treat each other better, and to recognize our planet as the fragile haven that it is" [9, p. 318], a cosmic consciousness—indeed, a "cosmic responsibility"—reflecting our true cosmic connectedness.

Just how rare are pale blue dots in the Cosmos? In 1961, one year after he initiated an (unsuccessful) attempt to detect radio signals from communicative extraterrestrial civilizations, the astronomer Frank Drake (1930–2022) formulated the eponymous equation used to estimate the number of communicative civilizations in the Galaxy. One of the seven terms in the equation is the longevity or lifetime factor L, the number of years such civilizations exist to send signals, a highly uncertain parameter, being tied as it is to sociopolitical (un)sensibilities. Other factors, such as the rate of star formation in the Galaxy today (a few solar masses per year), the fraction of stars having planets (nearly half of all stars, we now know, although in his time, Drake's guess was as good as any), the average number of planets capable of supporting life associated with such stars (likely nearly one, based on current exoplanet architectures), and the fraction of those planets that actually develop life (probably all, given the propensity for life to exist whenever given a chance), intelligence (unknown, but so is what we mean by "intelligence"), and communicative ability (certainly all that develop "intelligence"), are, with the exception of the "intelligence" factor, less dependent on sociological conditions and are thus less uncertain. During the Cold War, estimates for L were as low as 50–100 years; we've been living with nuclear weapons and atomic power for about 80 years now. Many pundits, Carl Sagan included, believe that once a civilization develops the technology for interstellar communication, it will also have the technology to self-destruct: Ernest Rutherford's "moon-shine"—power from the atom[4]—may yet be the death of us. Climate change, ecological collapse, artificial intelligence (AI)—and, some might add, people of a certain political persuasion—are just some of the other existential crises humanity faces that could be shared across the Galaxy by other "intelligent" civilizations, lowering realistic estimates for L.

Let's hope we survive. After all, we're all stardust, each of us children of the stars, sharing a common chemistry and cosmic connection.

And so, as we come to the end of our story, let's follow the American poet Walt Whitman, and leave the "learn'd astronomer" and wonder off "[i]n the mystical moist night air" to look "up in perfect silence at the stars."[5] "Remember to look up to the stars," the late cosmologist Stephen Hawking advised (Fig. 5.3; and, he continued, "if you're lucky enough to find love, remember it is there and don't throw it away"). "We are all in the gutter," one of the characters in the nineteenth-century Irish poet and playwright Oscar Wilde's *Lady Windermere's Fan* utters, "but some of us are looking at the stars." Even in antiquity, Plato, writing in Book 7 of *The Republic*, admits that "it is obvious to everyone, I think, that this study [of astronomy] compels the soul to look upward and leads it away from things here to higher things" (recall Fig. 2.4). The Roman poet Ovid found solace in the skyward anatomy of the human body, writing at the beginning of the Christian era in his *Meta-morphoses* that "While the other creatures on all fours look downwards, man was made to hold his head erect in majesty and see the sky and raise his eyes to the bright stars above." And in more modern times, the nineteenth-century post-Impressionist painter Vincent van Gogh admitted to his brother Theo that whenever he felt the "need for, shall I say the word—for religion" he would "go outside at night to paint the stars." And paint the stars he did. Figure 5.4, his 1889 *The Starry Night*, is one of my favorites and certainly one of the most recognizable paintings in Western art—and a fitting final image for our story of our starry cosmic connection.

---

[4] At a 1933 meeting of the British Association for the Advancement of Science, this "father of nuclear physics" warned that "The energy produced by the breaking down of the atom is a very poor kind of thing. Anyone who expects a source of power from the transformation of these atoms is talking moonshine." (Rutherford didn't live to witness Hiroshima and Nagasaki.)

[5] In the spirit of nineteenth-century Romanticism, Whitman is saying that the true way to understand nature is too *feel* it, to *experience* it, not scientifically but intuitively, not with the head but with the heart. Listening to the "learn'd astronomer," being "shown the charts and diagrams," made him "tired and sick."

Fig. 5.3 Stephen Hawking and friends, Amsterdam, June 2003. (*Photograph courtesy of the author's daughter, Brittany, shown here standing next to the esteemed scientist*)

Fig. 5.4 Vincent van Gogh's magnum opus, his 1889 *The Starry Night*, now in New York City's Museum of Modern Art (MoMA). Obviously impressed by the starry sky, the brightest "star" in the painting just to the right of the foreground cypress tree is the "morning star," the planet Venus, he described in a letter to his brother Theo. The whirly sky may reflect the painter's familiarity with Lord Kelvin's then-popular vortex theory of the atom, or perhaps he was influenced by the widely publicized sketches of the spiral structure of what became known as the Whirlpool galaxy (M51) discovered by William Parsons (1800–1867), the third Earl of Rosse, using what was the largest telescope in the world for nearly three-quarters of a century on the grounds of his castle at Parsonstown (now Birr) Ireland. (*Wikimedia Commons, public domain*)

But, alas, it's nearly too late to look up at the stars: light pollution hides most of the stars in the sky from most of us, destroying our most immediate and afferent connection with the Cosmos (and disrupting plant growth and animal migration, feeding, and breeding). We live most of our lives inside under a roof with a plethora of distractions instead of outside under the stars wrapped in awe and wonder. We have clocks and calendars to tell us time and season—and artificial lighting (thus the light pollution) to change night into day any day of the year as we see fit. We're more detached from the stars and cycles in the sky than ever before. And yet very few of us seem concerned by a loss our ancestors enjoyed ever since we first looked up and wondered what it's all about. But we can still wonder. And hope. We're good at that, "this quintessence of dust" such as we are [10].

Of all the sciences cultivated by mankind, astronomy is acknowledged to be, and undoubtedly is, the most sublime, the most interesting, and the most useful. For by knowledge derived from this science, not only the bulk of Earth is discovered… but our very faculties are enlarged with the grandeur of the ideas it conveys, our minds exalted above the low contracted prejudices of the vulgar.
     – James Ferguson, *Astronomy Explained upon Sir Isaac Newton's Principles*
(1757)

The heavens afford the most sublime subject of study which can be derived from science….
     – Scottish polymath Mary Somerville, *On the Connexion of the Physical Sciences* (1834, pp. 1–2)

We are stardust
Billion year old carbon
We are golden …
And we've got to get ourselves
Back to the garden.
     – Singer-songwriter Joni Mitchell, "Woodstock" (1969)

References

1 J. Gribbin, *Stardust* (The Penguin Group, Chatham, UK, 2000)
2 C. Sagan, *Cosmos* (Random House, New York, 1980)
3 M. Marder, "From Cracks to Atoms and Back Again," Phys. Today 77, 62–63 (2024) https://doi.org/10.1063/pt.ihuk.frks
4 A. Einstein, "What I Believe," Forum Century **84** (1930)
5 J. Al-Khalili, *The World According to Physics* (Princeton University Press, Princeton and Oxford, 2022)

6 G. B. Shaw, *The Adventures of the Black Girl in Her Search for God, in Collected Prefaces* (Hamlyn, London, 1963)

7 L. Eiseley, *The Immense Journey: An Imaginative Naturalist Explores the Mysteries of Man and Nature* (Random House, New York, 1957)

8 C. Sagan, *Pale Blue Dot: A Vision of the Human Future in Space* (Random House, New York, 1994)

9 J. Marchant, *The Human Cosmos: Civilization and the Stars* (Dutton, New York, 2020)

10 W. Shakespeare, *Hamlet*, Act II, Scene 2

Further Reading

Most introductory astronomy books cover, to a greater or lesser degree, the birth of the Universe and the life cycle of stars and the synthesis and dispersal of the chemical elements. The collection of articles written for a general audience by experts in the field, edited by Malkan and Zuckerman and referenced below, closely parallels the chapters here and can be consulted for more details. The very recent review article by Arcones and Thielemann, also referenced below, is an up-to-date and comprehensive resource on the origin of the elements, with minimal mathematics that can be skipped over, making it accessible to a wider audience. *The Physics of Stars* by A. C. Phillips (John Wiley & Sons, Chichester, UK, 2nd ed. 1999; orig. publ. 1994) is exactly as the title advertises and "sorts ... out with a minimum of fuss ... key aspects of stellar structure, evolution and nucleosynthesis." An excellent and concise astronomy and astrophysics resource at the advanced undergraduate level is Dan Maoz's *Astrophysics in a Nutshell* (Princeton University Press, Princeton & Oxford, 2007). Donald D. Clayton's *Principles of Stellar Evolution and Nucleosynthesis* (University of Chicago Press, Chicago & London, 1983; orig. publ. McGraw-Hill, New York, 1968), a popular graduate-level textbook and a rich resource for researchers, remains the standard work on stellar nucleosynthesis (Chapter 7 "Synthesis of the Heavy Elements" is especially relevant). David Arnett's *Supernovae and Nucleosynthesis: An Investigation of the History of Matter, from the Big Bang to the Present* (Princeton University Press, Princeton, 1996) is a thorough account of cosmic nucleosynthesis at an advanced level. I have found these sources indispensable in telling the story of our cosmic connection; other useful references appear below.

© The Editor(s) (if applicable) and The Author(s), under exclusive license to Springer Nature Switzerland AG 2024
R. Fleck, *We Are Stardust*, https://doi.org/10.1007/978-3-031-67275-0

References

1 J. Al-Khalili, *The World According to Physics*, (Princeton University Press, Princeton and Oxford, 2022)

2 A. Arcones, F.-K. Thielemann, "Origin of the Elements," Astron. Astrophys. Rev. **31**, 1 (2023). https://doi.org/10.1007/s00159-022-00146-x

3 P. Ball, *How Life Works: A User's Guide to the New Biology* (University of Chicago Press, Chicago, 2023)

4 J. Barrow, *The Origin of the Universe* (Basic Books, New York, 1994)

5 B. W. Carroll, D. A. Ostlie, *An Introduction to Modern Astrophysics*, 2nd edn. (Pearson Education Inc., San Francisco, 2007)

6 S. Chatterjee, *From Stardust to First Cells: The Origin and Evolution of Early Life* (Springer, Cham, 2023)

7 P. M. Dauber, R. A. Muller, *The Three Big Bangs: Comet Crashes, Exploding Stars, and the Creation of the Universe* (Addison-Wesley Pub. Co., Boston, 1995)

8 J. Eisberg, "Solar Science and Astrophysics," in *The Modern Physical and Mathematical Sciences (The Cambridge History of Science)*, vol. 5, ed. by M.J. Nye (Cambridge University Press, Cambridge, 2003), pp. 505–521

9 J. Farrell, *The Day without Yesterday: Lemaître, Einstein, and the Birth of Modern Cosmology* (Thunder's Mouth Press, Emeryville, CA, 2005)

10 A. V. Filippenko, "Stellar Explosions, Neutron Stars, and Black Holes," in *Origin and Evolution of the Universe: From Big Bang to Exobiology*, 2nd edn., ed. by M. A. Malkan, B. Zuckerman (World Scientific, Singapore, 2020), pp. 99–147

11 A. Frebel, T. C. Beers, "The Formation of the Heaviest Elements," Phys. Today **71**, 30–37 (2018). https://doi.org/10.1063/PT.3.3815

12 H.S. Goldberg, M. D. Scadron, *Physics of Stellar Evolution and Cosmology* (Gordon and Breach, New York, 1981)

13 D. Goldsmith, T. Owen, *The Search for Life in the Universe*, 3rd edn. (University Science Books, Sausalito, CA, 2001; orig. publ. Benjamin/Cummings, 1980)

14 J. Gribbin, *Stardust* (The Penguin Group, Chatham, UK, 2000)

15 E. Hadingham, *Early Man and the Cosmos* (University of Oklahoma Press, Norman, 1985)

16 S. Hawking, *A Brief History of Time: From the Big Bang to Black Holes* (Bantam Books, New York, 1988)

17 G. Holton, S. G. Brush, "Construction of the Elements and the Universe," in *Physics, the Human Adventure: From Copernicus to Einstein and Beyond*, Chapter 32 (Rutgers University Press, New Brunswick, NJ, 2001), pp. 499–516

18 K. Hufbauer, "Stellar Structure and Evolution, 1924–1939," J. Hist. Astron. **xxxvii**, 203–227 (2006). https://doi.org/10.1177/00218286060370

19 C. Impey, *The Living Cosmos: Our Search for Life in the Universe* (Random House, New York, 2007)

20 C. Impey, *Worlds Without End: Exoplanets, Habitability, and the Future of Humanity* (MIT Press, Cambridge, MA, and London, 2023)

21 H. Kragh, *Quantum Generations: A History of Physics in the Twentieth Century*, Chapter 23, "Cosmology and the Renaissance of Relativity" (Princeton University Press, Princeton, 1999)

22 H. Kragh, *Conceptions of Cosmos: From Myths to the Accelerating Universe* (Oxford University Press, Oxford, 2006)

23 H. Kragh, M. S. Longair (eds.), *The Oxford Handbook of the History of Modern Cosmology* (Oxford University Press, Oxford and New York, 2019)

24 K. R. Lang, *The Life and Death of Stars* (Cambridge University Press, Cambridge, 2013)

25 M. Longair, *The Cosmic Century: A History of Astrophysics and Cosmology* (Cambridge University Press, Cambridge and New York, 2006)

26 M. A. Malkan, B. Zuckerman, *Origin and Evolution of the Universe: From Big Bang to Exobiology*, 2nd edn. (World Scientific, Singapore, 2020)

27 J. North, *The Norton History of Astronomy and Cosmology* (W. W. Norton and Company, New York, 1994)

28 P. J. E. Peebles, *Cosmology's Century: An Inside History of our Modern Understanding of the Universe* (Princeton University Press, Princeton, 2020)

29 D. Prialnik, *An Introduction to the Theory of Stellar Structure and Evolution*, 2nd edn. (Cambridge University Press, Cambridge, 2010)

30 C. Sagan, *Carl Sagan's Cosmic Connection: An Extraterrestrial Perspective* (Cambridge University Press, Cambridge, 2000; orig. publ. Anchor Books, 1973)

31 C. Sagan, *Cosmos* (Random House, New York, 1980)

32 J. Silk, *The Big Bang*, 3rd edn. (W. H. Freeman & Co., New York, 2000)

33 V. Trimble, "The Origin and Evolution of the Chemical Elements," in *Origin and Evolution of the Universe: From Big Bang to Exobiology*, 2nd edn., ed. by M.A. Malkan, B. Zuckerman (World Scientific, Singapore, 2020), pp. 63–97

34 S. Weinberg, *The First Three Minutes: A Modern View of the Origin of the Universe* (Basic Books, New York, 1977; rev. ed. 1993)

Index

Printed in the United States
by Baker & Taylor Publisher Services